OXFORD LOGIC GUIDES

GENERAL EDITOR: DANA SCOTT

FORMAL
NUMBER THEORY
AND COMPUTABILITY

A WORKBOOK

By
ALEC FISHER

Lecturer in Philosophy, School of Social Studies
University of East Anglia

CLARENDON PRESS · OXFORD
1982

Oxford University Press, Walton Street, Oxford OX2 6DP

London Glasgow New York Toronto
Delhi Bombay Calcutta Madras Karachi
Kuala Lumpur Singapore Hong Kong Tokyo
Nairobi Dar es Salaam Cape Town
Melbourne Wellington
and associate companies in
Beirut Berlin Ibadan Mexico City

© Alec Fisher 1982

Published in the United States by
Oxford University Press, New York

British Library Cataloguing in Publication Data

Fisher, Alec
Formal number theory and computability:
a workbook. – (Oxford logic guides)
1. Logic, Symbolic and mathematical
I. Title
511.3   BC135
ISBN 0-19-853178-8

Typeset by Anne Joshua Associates, Oxford

Printed in Great Britain
By Billing and Sons Ltd.
London and Worcester

To Sarah, Daniel, Max, Susannah

PREFACE

I have enjoyed writing this book. Many people have helped and it is a pleasure to thank them now.

The Nuffield Foundation and the University of East Anglia generously supported work on an earlier draft of the text and enabled Rhona House to help me with many of the exercises and answers and also with some of the text. This help got the project off to a good start. When I ran into difficulties in Chapter 10, John Shepherdson disentangled me. Dana Scott, as editor of the series, gave me enormous help and encouragement. Judith Sparks and Muriel Parke patiently converted messy manuscript into tidy typescript and numerous students made improvements to earlier versions. My University and Oxford University Press have been helpful, generous, and patient throughout. But for my wife and children, to whom this book is affectionately dedicated, it would have appeared long ago!

Norwich 1981   A. E. F.

CONTENTS

INTRODUCTION

Since this text is meant to be used in a particular way it is important to read this introduction carefully first.

Although the book requires very little specific background in mathematics, logic, or computing, it is intended for students of mathematics, computing, or philosophy who have, roughly speaking, the mathematical sophistication of a first-year university mathematics undergraduate. Starting from this base it is intended as a *first* course in mathematical logic.

This is not only a first course in mathematical logic, it is also a short one. Students in the University of East Anglia cover the material presented here in nine weeks with three hours of classes per week. A standard short first course in mathematical logic would cover propositional and predicate logic. A course which included formal number theory would normally be much longer.

As a short first course in mathematical logic its contents are unorthodox. It delays very little over propositional and predicate logic and proceeds as quickly and directly as possible to proving Gödel's incompleteness results for elementary number theory and various related results. In fact the book is the author's answer to the problem: what can be most usefully achieved in a short course, starting from scratch with mathematically able students, *firstly* to show them the distinctive character of mathematical logic and prove some of its fundamental results *especially for those who will never do any more of the subject*, and *secondly* what can be done to attract the interest of those who might wish to go on? The material presented here is well worth mastering in itself and certainly satisfies the first requirement. It is unusual for students to get so far without being excited by the results and their implications, and also impressed by the power and novelty of the methods employed, so it satisfies the second requirement too. It is true that mathematical logic is now a huge subject and this course presents only a fragment of it, but it is an important fragment, which stands on its own and beckons invitingly. All this is less true of propositional and predicate logic.

Not only are the contents unorthodox but the method of exposition is unusual too. The text contains many hundreds of exercises and there is a large appendix containing model answers to nearly every one. The exercises are an integral part of the exposition and each one should be answered *as the student reaches it*. Most of them are simply comprehension exercises, checking that a previous part of the text has been understood, but some supply results without which the reader cannot proceed. Of course there is no harm in browsing through a chapter first, but when it is being properly worked, the exercises should be done as they are met. Answers should be written out carefully and *then* checked against model answers. Most of the exercises are intended to be easy so that the student is encouraged by success to go on. Some are more difficult of course. When the reader cannot solve a problem, he should use the 'Hint' if there is one, or even look at the beginning of the model solution. When he cannot solve a problem and cannot understand the model solution, that is the time to consult the teacher.

The structure of the course is clear from the list of contents but some points need further explanation. Chapters 1 and 3 are introductory. They provide the basic number theory and logic for what follows. Chapter 2 induces students to make explicit the mathematical and logical assumptions of Chapter 1: this is easier than it looks. The teacher could reasonably ask students to complete Chapters 1–3 before the course begins. Chapter 4 is important because it *motivates* much of what follows, especially the definition of *formal theory*. Unless the reader works carefully through Chapter 4 he may be bewildered by Chapters 5 and 6. These present the formal number theory, **N**, and various results provable within **N**. This theory is essentially that of S. C. Kleene (1952) (also in his 1967) to which Chapters 5 and 6 are heavily in debt. There is no ideal formal theory. The choice depends on one's purpose. This formal theory was chosen partly because it is well-worked, but also because students who want further reading are thereby given easy access to two excellent texts.

Although Chapter 7 contains an interesting example of a formal theory which can be shown to be complete, consistent, and decidable by finitary methods, it can be omitted by students who wish to take the direct route to Gödel's results or who are simply pressed for time.

Chapter 8 provides an easy introduction to computability. No previous knowledge is assumed. Many students now have some experience of computing so they find Chapters 8 and 9 very easy. The register machines of Shepherdson and Sturgis (1963) are 'natural', easy to handle, less cumbersome than Turing machines, and less abstract than recursive functions. Given the students for whom this book is intended they provide an ideal way to introduce computability. Those who do no more logic have had a useful introduction to computing. Those who go on can easily learn about recursive functions and their equivalence to R-computable functions (see Bell and Machover 1977).

Students may find Chapter 10 difficult, but this proof is easier than any other the author knows. Chapters 11 and 12 contain a large number of results related to the incompleteness and undecidability of formal number theory.

The test of this approach is whether it works. It has been tried for several years with students at the University of East Anglia and the evidence is that students can make surprisingly quick progress with relatively little help from the teacher.

Further reading is abundant and some is given in the Bibliography. Crossley *et al.* (1977), is a lively and discursive introduction to mathematical logic. Fraenkel (1953) is a good introduction to set theory, and the *Introduction* to Nagel and Newman (1959) is a useful supplement to the other parts of Chapter 4. There are several good texts which can be used to extend the material presented here, especially Boolos and Jeffrey (1974), Church (1956), Kleene (1952) and (1967), and Mendelson (1964). There is an excellent philosophical discussion of Gödel's incompleteness result in Dummett (1963).

In conclusion we note three conventions which are followed in this book.

Firstly, whether an expression belongs to informal number theory, or to a formal theory, or to 'metatheory' is indicated by the style of its individual *variables*. *Informal* number theory has *italic* variables, as in $a|m \cdot b + n \cdot c$ (p. 1); *formal* theories have *sans serif* variables, as in $\exists c(c' + a = b)$ (p. 45); and *metamathematical* expressions (see p. 44) have *roman* variables, as in $A(x) \supset (\exists x A(x) \supset A(x))$ (p. 53). Formal *numerals* are *bold italic*, thus, n (see p. 62).

Secondly, we omit brackets in accordance with standard mathematical practice wherever confusion will not occur: we insert them if it makes expressions easier to read. The convention followed in informal contexts is the same as that described on p. 45 for formal contexts, hence for example, $a \cdot x + b \cdot y$ means $(a \cdot x) + (b \cdot y)$ and $a \cdot b'$ means $a \cdot (b')$ and $1 + (i + 1)d$ means $1 + ((i + 1) \cdot d)$.

Finally, the symbol '∎' is used to mark the end of a definition, exercise, theorem, or proof.

PART I

The formalization
of number theory

1

PROVING SOME BASIC RESULTS

In this first chapter we present some of the key results of elementary number theory. The principal ones are: the Fundamental Theorem of Arithmetic (every number has a unique prime factorization); Euclid's Theorem (there are infinitely many primes); and the Chinese Remainder Theorem (the remainders upon dividing the terms of an arithmetic progression can be controlled). These facts are essential for all our subsequent work: not only are they representative of the core of the theory we wish to axiomatize, but we also need them in proving results *about* the formal axiomatic system. The proofs in this chapter are given in the usual informal style of mathematical textbooks, and the discussion of the logical background only begins in the next chapter.

Divisibility and primes

Among the integers ... $-3, -2, -1, 0, 1, 2, 3, ...$ we shall be particularly interested in the non-negative integers $0, 1, 2, 3, ...$ and we shall often call these the *natural numbers*. For the *sum* of numbers a and b we write '$a + b$'. For the *product* of a and b we write either '$a \cdot b$' or 'ab'.

DEFINITION 1.1.   For integers a, b, we shall say 'a *divides* b' if and only if there is an integer c, such that $a \cdot c = b$. We also say 'a *is a factor of* b' or 'b *is a multiple of* a' or 'b *is divisible by* a' and write the relationship as $a \mid b$. Its contradictory we write as $a \nmid b$. ■

THEOREM 1.1.   For all integers a, b, c, n, m, the following results hold:
  (i) $1 \mid a$;
  (ii) $a \mid a$;
  (iii) $a \mid 0$;
  (iv) if $a \mid b$ and $b \mid c$, then $a \mid c$;
  (v)   if $a \mid b$, then $a \cdot c \mid b \cdot c$;
  (vi) if $a \mid b$ and $a \mid c$, then $a \mid m \cdot b + n \cdot c$;
  (vii) if $a > 1$, then it is not the case that both $a \mid b$ and $a \mid b + 1$. ■

EXERCISE 1.1.   Prove Theorem 1.1 (i)–(vii). ■

Write out your answers to exercises very carefully because we will need them later for various purposes, including identifying which principles of logic are used in them. Prove them from the definitions of course, using the standard manipulations of algebra. Answers are provided in the Appendix. The symbol '■' that has been used at the end of the statements of these first definitions, theorems, and exercises is used to mark the end of the specifically numbered paragraphs. When a theorem is *proved*, the ■-mark will come at the end of the proof (where older books often used 'Q.E.D.').

Clearly $3|42$, $7|56$, $4|56$, and so on; but with 17 there are no (positive) divisors other than 1 and 17, and similarly for 13. Such numbers are of interest for many reasons.

DEFINITION 1.2.   A natural number p is *prime* if $p > 1$ and it has no positive divisors other than 1 and p. A number $n > 1$ which is not prime is called *composite*. ■

This definition makes 2 the least or first prime, and they proceed:

$$2, 3, 5, 7, 11, 13, 17, 19, 23, 29, 31, \ldots$$

The regularities and irregularities of this sequence remain a continuing puzzle. We discuss below why there are infinitely many primes, but the question of whether there are infinitely many pairs of primes p, $p + 2$ (e.g. 11, 13 and 29, 31) remains unsolved.

The Fundamental Theorem of Arithmetic

Composite numbers, note, are those that can be factored into smaller factors. How far can this factorization be carried? The Fundamental Theorem will show that a *unique* prime factorization is always possible; the proof will take several steps.

THEOREM 1.2.   Any natural number $n > 1$ can be factorized into prime factors whose product is n.

Proof.   We argue by complete induction. (The reader who is unfamiliar with proof by induction will find a discussion in Chapter 2.) The theorem is true for 2. In proving it for arbitrary n we assume the theorem true for all $m < n$. If n is prime there is nothing to prove. If n is composite there are natural numbers a, b such that $1 < a < n$, $1 < b < n$, and $a \cdot b = n$. But we know by the induction hypothesis that a, b are either primes or expressible as products of primes, and on substituting for them we get n expressed as a product of primes. ■

There are several possible ways of proving the Fundamental Theorem; because of our later interests we take the route via the Euclidean algorithm for finding a greatest of all the common divisors of natural numbers a, b.

DEFINITION 1.3.   The *greatest common divisor*, or GCD of natural numbers a, b, not both 0, is the largest number d which divides both a and b. We write this $d = G(a, b)$. ■
   Thus, $G(0, a) = a$. Clearly this definition could be extended to any finite set of numbers. Most people are familiar with the method of finding the GCD of a, b by factoring each into their prime factors, and then selecting the correct powers of common prime factors. An example should be attempted.

EXERCISE 1.2.   Find the GCD of 1404, 1560. ■

But clearly, one could not *use* such a procedure to prove the unique prime factorization theorem without circularity: The important thing about the Euclidean algorithm for present purposes is that it is a method for constructing the GCD of numbers a, b which in no way depends on factoring a, b into their prime factors. Before describing the algorithm we recall some simple facts about division.

THEOREM 1.3.   For natural numbers a, b, with $b \neq 0$, there exist *unique* integers q (the quotient) and r (the remainder) where $0 \leqslant r < b$ and

$$a = q \cdot b + r.$$

Proof. Because

$$0 < b < 2b < 3b < \ldots < nb < \ldots,$$

there must be a stage where the given number a satisfies:

$$q \cdot b \leqslant a < (q + 1) \cdot b.$$

This number q is unique. (Why?) We then let $r = a - (q \cdot b)$, and $r < b$ must hold. (Why?) ■

Ordinary long division of course always yields the q and r in a finite number of steps. The reader should check that the result also holds if either a or b is *negative* (where in the latter case we take $0 \leqslant r < -b$). For a further discussion see Stewart (1964), pp. 16–18. Theorem 1.3 is often referred to as the *Division Algorithm*, where the second word emphasizes that the quotient and remainder can actually be found.

The Euclidean Algorithm

The idea is to iterate the Division Algorithm. Take natural numbers a and b not both 0. If $b = 0$, stop and give a as *output*. If $b \neq 0$, perform the Division Algorithm to find q, r where $0 \leqslant r < b$ and $a = q \cdot b + r$. If $r = 0$, stop and give b as *output*. If $r \neq 0$, continue the process using the pair b, r. A sequence of equations will be found:

$$a = q \cdot b + r;$$
$$b = q_1 \cdot r + r_1;$$
$$r = q_2 \cdot b + r_2;$$
$$\vdots$$
$$r_{k-3} = q_{k-1} \cdot r_{k-2} + r_{k-1};$$
$$r_{k-2} = q_k \cdot r_{k-1} + r_k;$$
$$r_{k-1} = q_{k+1} \cdot r_k + 0;$$

where the successive remainders satisfy:

$$b > r > r_1 > r_2 > \ldots > r_{k-1} > r_k > 0.$$

Because the remainders are strictly decreasing, the process must stop with a remainder of 0. The *output* is the last non-zero remainder r_k. ∎

Two examples illustrate how the algorithm works.

(i)
$$1560 = 1 \cdot 1404 + 156$$
$$1404 = 9 \cdot 156 + 0.$$

(ii)
$$43 = 1 \cdot 30 + 13$$
$$30 = 2 \cdot 13 + 4$$
$$13 = 3 \cdot 4 + 1$$
$$4 = 4 \cdot 1 + 0.$$

THEOREM 1.4.   For any pair a, b of natural numbers, not both zero, the Euclidean Algorithm yields as output the GCD. Furthermore, we can write

$$G(a, b) = a \cdot x + b \cdot y$$

for suitable integers x, y.

Proof. Consider the algorithm as described for a pair a, b and call the output d. If $b = 0$, then $d = a = G(a, 0) = G(a, b)$. If $b \neq 0$, but $r = 0$, then

$d = b = G(q \cdot b, b) = G(a, b)$. If $r \neq 0$, we wish to show $d = r_k$, the last non-zero remainder, is the GCD.

Looking at the equations of the algorithm, we see that $d = r_k | r_{k-1}$. But then, also, $d | r_{k-2}$. Working *backwards* we find:

$$d | r_{k-3}, \ldots, d | r_2, d | r_1, d | r,$$

and by the same token, $d | b$ and $d | a$. This proves that d is *a* common divisor. If D is any other common divisor, then $D | r$. (Why?) Thus working *forwards* we have:

$$D | r_1, D | r_2, \ldots, D | r_{k-1}, D | r_k.$$

So $D | d$, and this proves d is indeed the GCD.

For the second part of the theorem, note that $r = a - (q \cdot b)$; that is r is a 'linear combination' of a and b of the form $a \cdot x + b \cdot y$ with $x = 1$ and $y = -q$. Substituting this expression for r into the second equation, we find r_1 is also a linear combination of a and b. Continuing down the line we finally prove that $r_k = d = G(a, b)$ is a linear combination of a and b also. ∎

THEOREM 1.5. If p is a prime and $p | ab$, then $p | a$ or $p | b$. ∎

EXERCISE 1.3. Prove Theorem 1.5.
(*Hint.* If $p \nmid b$, then $1 = G(p, b)$. Use Theorems 1.4 and 1.1.) ∎

COROLLARY 1.5. If p is a prime and $p | abc \ldots n$, then

(i) $p | a$ or $p | b$ or ... or $p | n$;

(ii)   if $a, b, c, \ldots, n$ are all primes, then $p = a$ or $p = b$ or ... $p = n$. ∎

EXERCISE 1.4. Prove Corollary 1.5. ∎

THEOREM 1.6 (*The Fundamental Theorem of Arithmetic*).   Every natural number $n > 1$ has a *unique* prime factorization of the form

$$n = p_1^{a_1} \cdot p_2^{a_2} \cdot \ldots \cdot p_k^{a_k},$$

where $k \geqslant 1$, each p_i is prime, each $a_i \geqslant 1$, and

$$p_1 < p_2 < p_3 < \ldots < p_k.$$

Proof. The representation of the theorem, following custom, is described as 'writing n in standard form'. Suppose n can be written in standard form in two different ways

$$n = p_1^{a_1} \cdot p_2^{a_2} \cdot \ldots \cdot p_k^{a_k} = q_1^{b_1} \cdot q_2^{b_2} \cdot \ldots \cdot q_m^{b_m} \qquad (1)$$

Clearly $p_1 | q_1^{b_1} \cdot q_2^{b_2} \cdot \ldots \cdot q_m^{b_m}$; so by Corollary 1.5 (ii), $p_1 = q_i$ for some i and

$q_i \geqslant q_1$. Thus $p_1 \geqslant q_1$. Similarly $q_1 | p_1^{a_1} \cdot p_2^{a_2} \cdot \ldots \cdot p_n^{a_n}$; so, again, $q_1 = p_j$ for some j and $p_j \geqslant p_1$. Thus $q_1 \geqslant p_1$, and therefore $p_1 = q_1$.

Now suppose $b_1 \geqslant a_1$. Then $p_1^{a_1} = q_1^{a_1}$ can be divided out of the equation (1) leaving

$$p_2^{a_2} \cdot p_3^{a_3} \cdot \ldots \cdot p_k^{a_k} = q_1^{b_1 - a_1} \cdot q_2^{b_2} \cdot \ldots \cdot q_m^{b_m}.$$

If $b_1 > a_1$ the prime q_1, by the same argument as before, must equal some $p_j, j \geqslant 2$ but since $q_1 = p_1 < p_j$ for all $j \geqslant 2$ we have a contradiction, therefore $b_1 = a_1$. The supposition that $a_1 \geqslant b_1$, leads by the same argument to the conclusion that $b_1 = a_1$. We can thus cancel p_1 (and q_1) from both sides of (1), and, by repeating the above two steps for each successive prime and its power, the theorem is proved. ∎

The infinitude of the primes

The Fundamental Theorem of Arithmetic is appropriately named, because so many developments in number theory depend upon it, but notice that it has been proved without any reference to the question, 'How many primes are there?' We now look briefly at this question.

It is easy to construct a table of primes up to a modest limit by the 'Sieve of Eratosthenes', which for finding large primes is cumbersome though theoretically usable. The method runs as follows: keep 2 for it is a prime, but eliminate all the other even numbers:

$$2 \quad 3 \quad \cancel{4} \quad 5 \quad \cancel{6} \quad 7 \quad \cancel{8} \quad 9 \quad \cancel{10} \quad 11 \quad \cancel{12} \quad 13 \quad \cancel{14} \quad 15 \quad \cancel{16} \quad 17.$$

The next number remaining is 3 and it must be a prime. (Why?) Keep it and cross out all multiples of 3:

$$2 \quad 3 \quad \cancel{4} \quad 5 \quad \cancel{6} \quad 7 \quad \cancel{8} \quad \cancel{9} \quad \cancel{10} \quad 11 \quad \cancel{12} \quad 13 \quad \cancel{14} \quad \cancel{15} \quad \cancel{16} \quad 17.$$

The next number remaining is 5 and it must be a prime. (Why?) Keep it and cross out all multiples of 5:

$$2 \quad \cancel{3} \quad \cancel{4} \quad 5 \quad \cancel{6} \quad 7 \quad \cancel{8} \quad \cancel{9} \quad \cancel{10} \quad 11 \quad \cancel{12} \quad 13 \quad \cancel{14} \quad \cancel{15} \quad \cancel{16} \quad 17.$$

But in this list there were *no* other multiples of 5 (why?) and all the other numbers remaining must be prime. (Why? Because if a number n is *not* prime it must have a prime divisor $p \leqslant \sqrt{n}$.) In this way a table of all primes up to 100 can be made very quickly.

Tables which have been constructed by this method (and other considerations) suggest that there may be infinitely many primes, but the next theorem may be surprising.

THEOREM 1.7.   For any natural number n, there is a block of *consecutive composite* numbers whose length exceeds n.

Proof. If the number of primes is finite the result is immediate. If the number of primes is infinite let p_n be the nth prime and consider the primes 2, 3, 5, ... p_n. Then all the numbers from 2 up to p_n are divisible by one of these primes; so if we let $q = 2 \cdot 3 \cdot 5 \cdot ... \cdot p_n$, then all of the $p_n - 1$ numbers

$$q + 2, q + 3, q + 4, ..., q + p_n.$$

are composite (by Theorem 1.1 (vi) etc.). ∎

It may seem remarkable, and perhaps even counter-intuitive, that there can be both huge intervals of composite numbers (even as long as the largest known prime, $2^{11213} - 1$, whose *decimal expansion* has 3376 digits!) and also infinitely many primes; but the latter is in fact easily proved. The best known proof was known to Euclid in the fourth century BC, *Elements* (Bk. vii, Prop. 30). This proof is given in our model solution in the Appendix.

THEOREM 1.8.   The number of primes is infinite. ∎

EXERCISE 1.5.   Prove Theorem 1.8.
    (*Hint.* If the number of primes is finite, multiply them together and add 1. What prime divisors can this large number have?) ∎

The Chinese Remainder Theorem

We conclude this section with a result which is very important for our later work (Theorem 1.11 below). But first we must note some more results concerning divisibility.

THEOREM 1.9.
    (i)   If a and b leave the same remainders when divided by c and $a > b$, then $c \,|\, (a - b)$.
    (ii)   If $c \,|\, a$ and $c \,|\, b$ and $a > b$, then $c \,|\, (a - b)$. ∎

EXERCISE 1.6. Prove Theorem 1.9. ∎

THEOREM 1.10.
    (i)   If p, q are relatively prime (i.e. have no common factors other than 1) and $p \,|\, a$ and $q \,|\, a$, then $p \cdot q \,|\, a$.
    (ii)   If $p \neq q$ are both prime and $p \,|\, a$ and $q \,|\, a$, then $p \cdot q \,|\, a$. ∎

EXERCISE 1.7.   Prove Theorem 1.10.
    (*Hint.* (i) use Theorem 1.4 above.) ∎

We know, by Theorem 1.3, that there exists a unique remainder r when an arbitrary number c is divided by $d > 0$. We now write $rm(c, d)$ for r so that we have

$$c = d \cdot q + rm(c, d)$$

and $rm(c, d) < d$. As c takes successively the values 0, 1, 2, 3, 4, ... the quantity $rm(c, d)$ takes cyclically d different values; namely:

$$0, 1, 2, \ldots, (d - 1).$$

For $n + 1$ different $ds, d_0, d_1, \ldots, d_n$, consider the $(n + 1)$-tuple of remainders:

$$(*) \qquad \langle rm(c, d_0), rm(c, d_1), \ldots, rm(c, d_n) \rangle.$$

Since the ith component of $(*)$ has d_i different values, it follows that, as c takes successive values we get *no more than* $d_0 \cdot d_1 \cdot \ldots \cdot d_n$ different $(n + 1)$-tuples of the form $(*)$. Note that two $(n + 1)$-tuples are said to be *equal* if and only if they are termwise equal; in other words sequences $\langle a_0, \ldots, a_n \rangle$ and $\langle b_0, \ldots, b_m \rangle$ are said to be *equal* if and only if $n = m$ and, for all $i \leq n$, we have $a_i = b_i$.

EXERCISE 1.8.   Are the following true?

 (i) $\langle 2, 3 \rangle = \langle 3, 2 \rangle$;

 (ii) $\langle 2, 0, 3 \rangle = \langle 2, 3 \rangle$;

 (iii) $\langle 2, 3, 3 \rangle = \langle 2, 3 \rangle$. ∎

Note also that in general for the special $(n + 1)$-tuple $(*)$ there will not necessarily be as many as $d_0 \cdot d_1 \cdot \ldots \cdot d_n$ distinct $(n + 1)$-tuples.

EXERCISE 1.9.
 (i)   Let $n = 1, d_0 = 2, d_1 = 6$ and verify that only six (*not* twelve) different couples $\langle rm(c, d_0), rm(c, d_1) \rangle$ are actually obtained.
 (ii) What is the case if $n = 1, d_0 = 3$ and $d_1 = 4$? ∎

We now state the condition under which there are *exactly* the maximal number of different $(n + 1)$-tuples.

THEOREM 1.11 (*The Chinese Remainder Theorem*).   If $d_0, d_1, \ldots, d_n$ are *relatively prime* (i.e. no two of them have a common factor other than 1), then, as c takes the successive values 0, 1, ..., $((d_0 \cdot d_1 \cdot \ldots \cdot d_n) - 1)$ exactly $d_0 \cdot d_1 \cdot \ldots \cdot d_n$ different $(n + 1)$-tuples $(*)$ are actually obtained.

Proof. Put $d_0 \cdot d_1 \cdot \ldots \cdot d_n = D$. If the theorem is false, then there exists c_1, c_2 such that

(1) $$0 \leq c_1 < c_2 < D$$

and such that c_1 and c_2 yield identical $(n + 1)$-tuples (*). From (1) it follows that $c_2 - c_1$ is positive and less than D. And since c_2 and c_1 leave the same remainders when divided by $d_0, d_1, \ldots, d_n$, it follows from Theorem 1.9 (i) that the difference $c_2 - c_1$ must be divisible by each of the numbers $d_0, d_1, \ldots, d_n$. But since the d_i are relatively prime, $c_2 - c_1$ must be divisible by their product D (Theorem 1.10 (i) used repeatedly). This is a contradiction since no positive number less than D can be divisible by D. ∎

Gödel's β-function

This result gives us a way of handling finite sequences of natural numbers employing a suggestion due to Gödel. Gödel's so-called β-function is defined as follows:

$$\beta(c, d, i) = \text{rm}(c, 1 + (i + 1)d).$$

By a suitable choice of d, the Chinese Remainder Theorem can now be invoked.

THEOREM 1.12.   For any finite sequence of natural numbers $a_0, a_1, \ldots, a_n$, we can find natural numbers c, d such that for $0 \leqslant i \leqslant n$

$$\beta(c, d, i) = a_i.$$

Proof. For the sequence $a_0, a_1, \ldots, a_n$ let s be the greatest of $n, a_0, a_1, \ldots, a_n$ and let $d = s!$ (Where $s! = s \cdot (s - 1) \cdot (s - 2) \cdot \ldots \cdot 1$.) Notice that

$$a_0 < 1 + d, \; a_1 < 1 + 2d, \ldots, a_n < 1 + (n + 1)d$$

because $a_i \leqslant s \leqslant d < 1 + (i + 1)d$. Therefore each a_i is a possible value for $\text{rm}(c, 1 + (i + 1)d)$. Now, by Theorem 1.11 we know that — if all the numbers $1 + (i + 1)d$ are relatively prime — we can find a c less than their product such that for $0 \leqslant i \leqslant n$

$$a_i = \text{rm}(c, 1 + (i + 1)d).$$

To show that the numbers in question are relatively prime, suppose they are not; that is, suppose that $1 + jd$ and $1 + kd$, with $1 \leqslant j < k \leqslant n + 1$, possess a common factor. Then there is a prime number p dividing both $1 + kd$. But then p divides their difference $(k - j)d$ (by Theorem 1.9 (ii)). Now p cannot divide d because, by hypothesis, it divides $jd + 1$ (Theorem 1.1). But neither can p divide $k - j$, because then $p \leqslant k - j \leqslant n \leqslant s$; therefore, it would follow that p divided $s! = d$, something which we already know is impossible. But if a prime number p divides neither d nor $k - j$ it cannot divide their product $(k - j)d$ (contrapositive of Theorem 1.5). Thus all the numbers $1 + (i + 1)d$ for $0 \leqslant i \leqslant n$ must be relatively prime. ∎

The question of elementary methods

Since every prime $p_i > 2$ is odd, they all fall into one of the following arithmetic progressions:

(a) 1, 5, 9, 13, 17,... (numbers of the form $4n + 1$);

(b) 3, 7, 11, 15, 19,... (numbers of the form $4n + 3$).

It is not too difficult to prove that there are infinitely many primes in *each* progression (though different kinds of proof are needed). The same conclusion can be derived also for the series of numbers of the form $6n - 1$, or $6n + 1$, $8n + 5$ and various others.

These results suggest a generalization. Arithmetic progressions consist of all numbers of the form $an + b$, for fixed a, b and $n = 0, 1, 2, 3,....$ Now, if a and b have a common factor, all members of the progression have this factor, so it can contain at most one prime (its least number). But if a, b have no common factor, if they are relatively prime, what then? Attempts to settle this general question met with little success, and the proofs which worked for the above progressions simply would not generalize. Not until 1837 was Dirichlet able to find his proof.

THEOREM 1.13.. If $a \geq 0$ and a, b are relatively prime then there are infinitely many primes of the form $an + b$. ∎

Dirichlet's proof is enormously difficult. We mention it as the first example of a proof which used concepts and truths about *real* numbers (especially, functions of a continuous variable, limits, infinite series, etc.) that is to say, *analytic* methods, to prove a purely *arithmetic* theorem. In 1949 A. Selberg proved this theorem by purely arithmetic means, but the proof is also far too difficult even to discuss in outline here.

This example shows that theorems may have a very elementary formulation but highly complex proofs. However, in the end a proof of 1.13 was found which was elementary. The problem therefore arises whether a theorem with an elementary statement always has in principle an elementary proof. We return to this question in Part II of the book.

2

ELICITING ASSUMPTIONS

In Chapter 1 some basic results of number theory were presented and proved in the standard mathematical way. These proofs take for granted many things which we must now make explicit: conventions about the use of variables, logical steps the reader probably never thought about before, and truths about numbers which are usually thought too obvious to need mentioning. We shall return often to the discussion of why this explicitness is necessary.

Eliciting assumptions: a first example

Let's begin by looking at the proof of Theorem 1.1 (i) which states that $1 \mid a$ (for any integer a). To prove this from the definition, we need to show that there is an integer c such that $1 \cdot c = a$. Clearly, since $1 \cdot a = a$, *there is* a c such that $1 \cdot c = a$, so $1 \mid a$. The role of the variable a in the proof may be described thus: a stands for any arbitrarily chosen integer, the same one in all occurrences throughout the proof.

The three steps in the proof are these:

Step 1. Clearly, the starting point or initial premiss, is that $1 \cdot a = a$ (for any integer a). No one doubts that this is true, and it is a natural candidate to become an axiom if we are seeking to systematize our knowledge of the integers (or of the natural numbers) into an axiomatic, deductive system. We can accept it because it is very simple and obviously true, and it is not easy to see what more basic truths might serve to justify it.

Step 2. From this basis we have no difficulty in taking the logical step to 'there is a c such that $1 \cdot c = a$'. If $1 \cdot a = a$ (for any integer a) then, clearly, there is a c such that $1 \cdot c = a$, namely a itself! In contrast to the number-theoretic truth with which the proof begins, the inference used in Step 2 is not of a kind peculiar to number theory:

... if everything in some class bears some relation, R, to itself, then for any member of that class there is something which stands in the relation R to that initial arbitrary member ...

But this is too cumbersome. English is no more suitable for logic than it is for mathematics, so we list now, for convenience, the notation we shall use for logical operations (or connectives):

'A ⊃ B' reads 'if A, then B' or 'A implies B'.

'A ∼ B' reads 'A if and only if B' or 'A is equivalent to B'.

'A & B' reads 'A and B'.

'A ∨ B' reads 'A or B'

'¬A' reads 'not A' or 'it is not the case that A'.

'∃xA(x)' reads 'for some x, A(x)' or 'there is an x which has the property A'.

'∀xA(x)' reads 'for some x, A(x)' or 'every x has the property A'.

Using this logical notation the logical principle used in Step 2 may be expressed as

$$R(a, a) \supset \exists b R(a, b).$$

This principle is obviously true for any class of objects and any relation R, so it is not a truth of number theory particularly, nor of physics, nor of biology, but of *logic*.

Step 3. There is 'one' last step in the proof. From the definition, $1 \mid a$ means the same as $\exists c(1 \cdot c = a)$, i.e. $1 \mid a$ *is true* if and only if $\exists c(1 \cdot c = a)$ *is true*. Clearly then, since we already know that $\exists c(1 \cdot c = a)$ we can conclude that $1 \mid a$. Letting A, B stand for the mathematical propositions $1 \mid a$ and $\exists c(1 \cdot c = a)$ respectively, we can break down this one move into the following steps:

(i)   A ∼ B   by definition;

(ii) (A ∼ B) ⊃ (B ⊃ A)   general truth about propositions;

(iii) B ⊃ A   from (i) and (ii) by *modus ponens*;

(iv) B   proved in previous *Step 2*;

(v)   A   from (iii) and (iv) by *modus ponens*.

In steps (iii) and (v) '*modus ponens*' (MP) is the name given to the obviously true principle that if A implies B, and A is true, so is B. This is so obvious the reader probably never gave it a thought before and certainly cannot doubt it! The same goes for step (ii).

In conclusion, in proving this very simple theorem, we assumed that the variable *a* was used in a certain way, we assumed one truth about integers (namely, $1 \cdot a = a$), one principle of predicate logic ($R(a, a) \supset \exists b R(a, b)$), one principle of *propositional* logic ((A ∼ B) ⊃ (B ⊃ A)), one rule of inference (*modus ponens*), and the legitimacy of our definition of $a \mid b$. All these assumptions *are* fully justified! At this stage we only wish to draw attention to the matters we took for granted in constructing our original proof.

Part of the motivation for drawing out implicit assumptions is to avoid error. As a matter of fact the history of mathematics is littered with 'proofs' which turned out later to be incorrect or to depend upon questionable implicit assumptions. This would have remained a matter for *mathematical* correction had not serious logical paradoxes been discovered at the end of the nineteenth century, which showed that some of the most deeply rooted assumptions of mathematical practice gave rise to *contradictions*! However, there is no suggestion that any of the principles we have *so far* used give rise to error, because they are far too elementary. There are much more positive reasons for making our underlying logic explicit which provide the main motivation for this exercise and which will emerge from our discussions.

EXERCISE 2.1.   Carry out parallel analyses for your proofs of Theorem 1.1 (ii) and (iii). ■

A 'direct' proof

Now look at Theorem 1.1 (iv), which states that

$$a \mid b \ \& \ b \mid c \supset a \mid c,$$

for any integers a, b, c. The gist of the proof is this: assume $a \mid b$ and $b \mid c$. We know by definition that there is an n and an m such that $a \cdot n = b$ and $b \cdot m = c$. Substituting for b this gives $a \cdot n \cdot m = c$, so $\exists k(a \cdot k = c)$, therefore $a \mid c$. This very simple argument is, logically, surprisingly complex and can be broken down into several more basic steps. We shall deliberately skate over some of the complications at this stage: the reader will be able to appreciate them and deal with them by the time he encounters the formal proof of this result in Chapter 6.

In common usage among mathematicians there are two standard ways of attempting to prove a proposition of the form

'if A(is true), then B(is true)'

or, equivalently, 'A implies B'. The first method, known as 'direct proof', assumes A true, and, on that assumption, tries to show that B is true. If successful we conclude that A ⊃ B. The second method, known as 'indirect proof' or '*reductio ad absurdum*', assumes both that A is true and that B is *false*, and on these assumptions tries to derive a *contradiction*. If successful, we conclude that if A is true B *cannot* be false (because *that* gives a contradiction); hence, if A(is true) then B(is true), or A ⊃ B. We shall encounter many examples of indirect proof soon, so for the moment we shall only consider proving Theorem 1.1 (iv) by the direct method used in our proof above.

The proof begins with the *assumption* that $a \mid b \ \& \ b \mid c$. We first note that in this proof the variables are taken to stand for arbitrary integers a, b, c which satisfy the condition $a \mid b \ \& \ b \mid c$. If the conjunction $a \mid b \ \& \ b \mid c$ is true then so are its 'conjuncts'; so, by the trivial principles that

$$(A \ \& \ B) \supset A \text{ and } (A \ \& \ B) \supset B$$

(and *modus ponens*), we can conclude separately that $a \mid b$ and that $b \mid c$. We can apply the definition of divisibility and an obvious variation on the sequence (i) ... (v) in *Step 3* above to show that

$$\exists n(a \cdot n = b) \text{ and } \exists m(b \cdot m = c).$$

EXERCISE 2.2.   Produce the 'obvious variation'. ∎

Having reached this point the standard mathematical way of dealing with the quantifiers $\exists n$ and $\exists m$ is to say 'let $a \cdot n = b$ and let $b \cdot m = c$'. So we shall do the same: we *assume* both of these, the variables $a, b, c, n,$ and m all standing for arbitrary integers which satisfy the conditions $a \cdot n = b$ and $b \cdot m = c$. If these assumptions imply $a \mid c$, because they are introduced only as a device for dealing with the quantifier, we conclude that $\exists n(a \cdot n = b)$ and $\exists m(b \cdot m = c)$ do also.

Now, using $a \cdot n = b$ and $b \cdot m = c$ *and* some principle of identity (e.g. that if $r = s$, then what is true of r is true of s), we can conclude that $(a \cdot n) \cdot m = c$. More specifically, using

$$a \cdot n = b \text{ and } b \cdot m = c$$

plus the principle that

$$r = s \supset r \cdot t = s \cdot t$$

plus the 'transitivity of identity',

$$r = s \supset (s = t \supset r = t),$$

we can conclude that

$$(a \cdot n) \cdot m = c.$$

Having established that last equation, we assumed the 'associativity' of multiplication for the natural numbers:

$$(a \cdot n) \cdot m = a \cdot (n \cdot m).$$

And from this we concluded without difficulty the existence of a number $k(= n \cdot m)$ such that $a \cdot k = c$. Thus, by the definition again, it follows that $a \mid c$.

To sum up, in proving this simple-minded result we used variables in the way described, two further principles of 'propositional' logic besides some of the logical principles already mentioned ((A & B) ⊃ A and (A & B) ⊃ B), two principles about identity ($r = s \supset r \cdot t = s \cdot t$ and $r = s \supset (s = t \supset r = t)$), a number theoretic principle (($a \cdot n) \cdot m = a \cdot (n \cdot m)$), and finally, two principles of 'predicate' logic (if some predicate P(a) is true for a particular number k, then $\exists a$P(a) is true, and a 'device' for deriving results from existentially quantified propositions – propositions of the form $\exists a$P(a)).

Although the last mentioned device is familiar to mathematicians, its general formulation is perhaps a little tricky. We shall put it like this: let C be a proposition which *does not contain n*, then if, for an arbitrary number n, $P(n) \supset C$, it follows that $(\exists a P(a)) \supset C$. The argument is justified because in proving C nothing about n was used except that it satisfied the assumption $P(n)$.

At the risk of repetition, but for ease of reference later, this proof may be set out in a linear form showing the justification for each step.

A model proof

1. $a \mid b \& b \mid c$ assumption.

2. $a \mid b$ by 1, by (A & B) $\supset$ A and *modus ponens* (MP).

3. $b \mid c$ by 1, by (A & B) $\supset$ B and MP.

4. $a \mid b \sim \exists n(a \cdot n = b)$ definition.

5. $\exists n(a \cdot n = b)$ by 2 and 4 by variation on (i)–(v) in *Step 3* above.

6. $b \mid c \sim \exists m(b \cdot m = c)$ definition.

7. $\exists m(b \cdot m = c)$ 3 and 6 by variation on (i)–(v) in *Step 3* above.

8. $a \cdot n = b$ assumption.

9. $b \cdot m = c$ assumption.

10. $(a \cdot n) \cdot m = b \cdot m$ by 8, $r = s \supset r \cdot t = s \cdot t$, and MP.

11. $(a \cdot n) \cdot m = c$ by 9, 10, $r = s \supset (s = t \supset r = t)$, and MP twice.

12. $a \cdot (n \cdot m) = c$ by 11, $(a \cdot n) \cdot m = a \cdot (n \cdot m)$,
                                $r = s \supset (s = t \supset r = t)$, and MP twice.

13. $\exists d(a \cdot d = c)$ by 12 and $P(k) \supset \exists d P(d)$.

14. $a \mid c$ by 13, definition and *Step 3*.

Again, all these further principles we have used are sound, and they are clearly the kind of basic truths we must elicit if we are to axiomatize number theory and its logic.

EXERCISE 2.3.   Carry out parallel analyses for your proofs of Theorem 1.1 (v) and (vi). ■

Reductio ad absurdum

We now look at Theorem 1.1 (vii), which states that

$$a > 1 \supset \neg(a \mid b \& a \mid b + 1).$$

To prove this, assume that $a > 1$ and that $(a \mid b \& a \mid b + 1)$; this means there exist n, m such that $a \cdot n = b$ and $a \cdot m = b + 1$. Since $b < b + 1$, we find $a \cdot n < a \cdot m$, so $n < m$. But then, there is an $x \geqslant 1$ such that $n + x = m$, so

$$b + 1 = a \cdot m = (a \cdot n) + (a \cdot x) = b + (a \cdot x).$$

Thus $1 = a \cdot x$ which contradicts the assumption that $a > 1$.

Without going into so much detail as previously we again note our assumptions in this proof. The pattern of the proof is that of a *reductio ad absurdum*; we assume A and B and derive a contradiction from their joint assumption, so we conclude that if A is true B must be *false*, in symbols A ⊃ ¬B. Since this method of proof is probably more familiar to mathematicians than any other, we shall take it for granted and discuss it no further here, except to symbolize the two forms of inference:

from ¬(A & B), to conclude (A ⊃ ¬B); and

from ¬(A & ¬B), to conclude (A ⊃ B).

It is not difficult to see that if the proof of 1.1 (vii) is spelled out in full detail it uses several logical principles already mentioned. It also rests on several number-theoretic truths:

$$a \cdot n < a \cdot m \supset n < m;$$

$$a + n = b + n \supset a = b;$$

$$a \cdot (m + n) = (a \cdot m) + (a \cdot n).$$

There are also others, some already mentioned; the details are left to the reader.

Mathematical induction

So far we have allowed variables to convey *generality*, as in $a \mid b \,\&\, b \mid c \supset a \mid c$, reminding the reader that they are playing this role by sometimes adding 'for any integers a, b, c'. In doing this we have simply been following standard mathematical practice — and making it explicit. However, in what follows we shall venture into more difficult territory; there we shall find it helpful to use the universal quantifier, $\forall a$, which reads 'for any a' or 'for all a'. Although it might have obscured things for many readers to be told that in Theorem 1.1 (iv) we proved

$$\forall a \forall b \forall c (a \mid b \,\&\, b \mid c \supset a \mid c),$$

when we discuss *mathematical induction*, the use of the universal quantifier will make the argument much clearer. As before, we shall omit it where there is no danger of confusion and it seems natural to do so, but where clarity demands we shall use it.

It is in the proof of Theorem 1.2 that we find mathematical induction employed for the first time. This method of proving a result for *all* the natural numbers is used constantly in number theory in various different forms. A common form is the passage from n to $n + 1$. For a mathematical property

A(x), suppose we have a proof that A(0) is true, plus a proof that if A(x) is true so is A($x + 1$); it can then be concluded that $\forall x$A(x). In our logical notation, writing x' as short for $x + 1$ we may express this as follows:

(I) $\qquad\qquad$ (A(0) & $\forall x$(A(x) $\supset$ A(x'))) $\supset \forall x$A(x).

But in the proof of Theorem 1.2, induction occurs in the form often called 'complete induction': if, for any natural number x, A(x) follows from the fact that all the predecessors y of x have the property A(y), then *every* number has the property. In our notation, making particular use of $\forall$, we write

(II) $\qquad\qquad \forall x(\forall y(y < x \supset$ A(y)) $\supset$ A(x)) $\supset \forall x$A(x).

Note that the placing of the quantifiers is very important to the reading of the statement. The first '$\forall x$' governs a different use of the variable from the second − it is as if a deep breath is needed between the two parts of the sentence. The same goes for formulation (I) as a matter of fact.

It is possible to prove that these two principles are equivalent, so that (given the rest of number theory) (I) implies (II) and (II) implies (I). Since the proof that (I) implies (II) is rather complicated, we omit it here, though its formal equivalent is proved in Chapter 6. The proof that (II) implies (I) is as follows: Assume that

$$A(0) \text{ \& } \forall x(A(x) \supset A(x')).$$

There are two cases, for an arbitrary x, either $x = 0$ or $x > 0$. By our initial assumption A(0), so, by the principle B $\supset$ (A $\supset$ B), we have

(i) $\qquad\qquad \forall y(y < 0 \supset$ A(y)) $\supset$ A(0).

Thus

(ii) $\qquad\qquad x = 0 \supset [\forall y(y < x \supset$ A(y)) $\supset$ A(x)].

If $x > 0$, let $x = z'$ and assume $\forall y(y < z' \supset$ A(y)). Since $z < z'$, and since A(z) and, by our initial assumption,

(iii) $\qquad\qquad\qquad$ A(z) $\supset$ A(z'),

it follows that A(z') holds. Hence,

(iv) $\qquad\qquad \forall y(y < z' \supset$ A(y)) $\supset$ A(z').

Thus $x = z' \supset \forall y(y < x \supset$ A(y)) $\supset$ A(x), so

(v) $\qquad\qquad x > 0 \supset [\forall y(y < x \supset$ A(y)) $\supset$ A(x)].

Using Proof by Cases and introducing the universal quantifier $\forall x$, we get

(vi) $\qquad\qquad \forall x(\forall y(y < x \supset$ A(y)) $\supset$ A(x)).

Hence, by (II), $\forall x$ A(x) does indeed follow.

Notice that, at (ii) we use the identity principle (if $a = b$ then $P(a) \supset P(b)$); at (iii) we 'eliminate' the universal quantifier (clearly, if every x has the property $A(x)$, any arbitrary number z has); at (iv) we 'introduce' $\supset$ (if from the assumption A we can derive B, clearly $A \supset B$); and finally, at (v) we 'introduce' the universal quantifier (clearly, if an arbitrarily chosen number x can be proved to have the property $A(x)$, then every number has it; so $\forall x A(x)$).

EXERCISE 2.4.   Write out the proof in linear form as in the earlier example giving the justification for each step. ■

Returning after that digression to the proof of Theorem 1.2, we find that complete induction is used to show that every number $n > 1$ can be factored into primes whose product is n: call this property $F(x)$. To establish $\forall x F(x)$ we first have to prove that

$$\forall y (y < x \supset F(y)) \supset F(x),$$

for arbitrary x, then by $\forall$-introduction, complete induction, and *modus ponens* we have the result. To prove the displayed formula, note that there are two cases, $x = 2$ or $x > 2$, the first being trivial.

EXERCISE 2.5.   Why do we have to consider two cases? ■

Excluded middle and proof by cases

In the proof of the second case of Theorem 1.2 we use two more principles not so far encountered. First we assume that either a number is prime or it is not:

$$\Pr(n) \vee \neg \Pr(n).$$

This is an instance of the so-called 'Law of the excluded middle'. (Again this is a principle so familiar in mathematics that it can hardly be doubted. Just as in the proof of Theorem 1.2, given in Chapter 1, mathematicians use it constantly and rarely bother to even say they are doing so. But if we are to list our underlying assumptions we must include this one too.) The proof then becomes a proof by cases: it is shown first that $\Pr(n) \supset F(n)$, then it is shown that $\neg \Pr(n) \supset F(n)$. Since by excluded middle these are the only possible cases, we conclude $F(n)$ for any $n > 2$. The general form of a proof by cases can be written

(*) $[(A \supset C) \,\&\, (B \supset C)] \supset [(A \vee B) \supset C]$.

The proof that $\Pr(n) \supset F(n)$ is trivial. The proof that $\neg \Pr(n) \supset F(n)$ is longer, however. If $\neg \Pr(n)$ then, by some simple steps from the definition of $\Pr(n)$, it follows that

$$\exists a\ \exists b\,(1 < a,\ b < n\ \&\ a \cdot b = n).$$

But, by the induction hypothesis $1 < y < n \supset F(y)$, so $F(a)$ and $F(b)$. Now, if a can be factored into primes whose product is a and similarly for b, and $n = a \cdot b$ clearly, by substituting their factors for a and b we shall find n similarly factored into primes whose product is n. That is to say, by our understanding of the meaning of $F(x)$, we accept

$$F(a)\ \&\ F(b) \supset F(a \cdot b).$$

So $\forall y(y < n \supset F(y)) \supset F(n)$, for non-prime n.

The proof is now concluded by considering an arbitrary $n > 2$ using proof by cases. Since $\Pr(n) \supset F(n)$, $\neg\Pr(n) \supset F(n)$ and $\Pr(n) \vee \neg\Pr(n)$, therefore, by (*) and *modus ponens*, we have $F(n)$. Then, using proof by cases again, on the cases $x = 2$ or $x > 2$ we have $\forall y(y < n \supset F(y)) \supset F(n)$ for arbitrary n. So by the logical principle that if property $A(n)$ is provable for an arbitrarily chosen number it is true for all numbers (often called 'universal generalization') we conclude finally that

$$\forall x(\forall y(y < x \supset F(y)) \supset F(x)).$$

By complete induction and *modus ponens* we have $\forall x F(x)$, which is what we wanted to prove.

To summarize, this proof uses mathematical induction, the Law of excluded middle, Proof by cases, and Universal generalization (among other things) all of which we take for granted commonly in mathematics.

EXERCISE 2.6.   Carry out a similar analysis for the proof of Theorem 1.5, which is also a proof by cases. ■

Summary of principles

In the preceding analyses we sought to do a number of things: to show very simply how the axiomatizing of a body of knowledge (theory) actually proceeds, to elicit certain truth about numbers which might serve as 'foundations' of the subject, and to articulate the logical principles we employ in mathematical argument without ever mentioning them. We review here some of the specific facts extracted from the examples we discussed.

Number-theoretic principles

$1 \cdot a = a$

$(a \cdot n) \cdot m = a \cdot (n \cdot m)$ (Associativity of times)

$a \cdot n < a \cdot m \supset n < m$

$a + n = b + n \supset a = b$

$$a \cdot (m + n) = (a \cdot m) + (a \cdot n)$$ (Distributivity)

$$a < a'$$

$$(A(0) \mathbin{\&} \forall x(A(x) \to A(x'))) \supset \forall x A(x)$$ (Induction)

$$\forall x(\forall y(y < x \supset A(y)) \supset A(x)) \supset \forall x Ax$$ (Complete induction)

Identity principles

$$r = s \supset r \cdot t = s \cdot t$$

$$r = s \supset (r = t \supset s = t)$$ (Transitivity of identity)

$$r = s \supset (P(r) \supset P(s))$$

Logical principles

$$R(a, a) \supset \exists b R(a, b)$$

If $P(k)$ for a particular number k, then $\exists a P(a)$ is true.

If $P(n)$ for an arbitrary number n, then $\forall a P(a)$ is true.

$$(A \sim B) \supset (B \supset A)$$

If $A \supset B$ and A, then B   (*Modus ponens*)

If from assumption A we can derive B, then $A \supset B$

If A and B imply a contradiction, then $A \supset \neg B$ (*Reductio ad absurdum*)

$$(A \mathbin{\&} B) \supset A$$

$$(A \mathbin{\&} B) \supset B$$

$$A \vee \neg A$$ (Excluded middle)

$$[(A \supset C) \mathbin{\&} (B \supset C)] \supset [(A \vee B) \supset C]$$ (Proof by cases)

$$B \supset (A \supset B)$$

If $\forall a P(a)$, then $P(k)$ for any number k.

If C is a proposition not containing n and if for an arbitrary number n $P(n) \supset C$, then $(\exists a P(a)) \supset C$.

Clearly, having analysed our theory thus far, we could continue with other proofs from Chapter 1 looking for other assumptions not yet mentioned; we could also look again at, say, the number-theoretic principles so far extracted and see if they too could be still further simplified and systematized by finding other, still more basic assumptions from which ours can be derived. Just such exercises have been carried out by previous workers in this field, and we shall shortly take advantage of all their efforts by offering an axiomatic system for elementary number theory with its underlying logic (both 'propositional' and 'predicate') and identity principles. This formalization has been arrived at by extended applications of just the methods we have shown above, and we shall have to discuss carefully how close it comes to being a *complete* axiomatization of elementary number theory (in the sense that every truth of the theory can be proved within it).

3

TESTING THE SOUNDNESS
OF LOGICAL ASSUMPTIONS

In the previous chapter we claimed that the principle

$$R(a, a) \supset \exists b R(a, b)$$

was 'obviously true for every class of objects and for any relation R'. We also 'took it for granted' that proof by cases was logically sound,

$$[(A \supset C) \,\&\, (B \supset C)] \supset [(A \lor B) \supset C],$$

that the Law of excluded middle,

$$A \lor \neg A$$

could 'hardly be doubted', and similarly with various other logical principles. Although we were confident of our judgement in these cases, would we have found it so easy to judge the soundness of the following principles?

  (i) $(A \supset \neg A) \supset \neg A$;

  (ii) $[(A \supset B) \supset A] \supset A$;

  (iii) $\forall a \,\exists b\, R(a, b) \supset \exists b \,\forall a\, R(a, b)$.

Is there a test for soundness?

We now digress from our main course to ask whether there is some test, some procedure we can apply, for deciding whether a putative logical principle is sound. In the case of principles of *predicate* logic we shall discover, in Chapter 12, that *there is no one test* which will enable us to decide whether an arbitrary principle is sound. We can however often settle the matter, for example (iii) above is easily shown to be unsound. But not *always* will matters be so simple.

EXERCISE 3.1.   Show that $\forall a \,\exists b\, R(a, b) \supset \exists b \,\forall a\, R(a, b)$ is unsound. ■

For principles of propositional logic on the other hand, there is a very neat decision procedure which accords with classical mathematical intuitions and which anyone interested in logic should know. We explain it briefly now.

Constructing truth tables

We lay down from the start this important assumption.

Any mathematical proposition P is either true or false (but not both). (*)

The negation of P (written as usual as ¬P) is false if P is true and true if P is false. So clearly, either P is true or ¬P is true, which is enough to guarantee that P ∨ ¬P is always true whatever P is. Given our initial assumption (*) we can use it as a reliable logical principle (excluded middle).

Again, consider the principle of *reductio ad absurdum*, which can be written as:

$$(P \supset (Q \mathbin{\&} \neg Q)) \supset \neg P.$$

If P implies a contradiction ((Q & ¬Q) must be false), P cannot be true, so it must be false. Notice that we argue these examples in general terms, for any propositions P, Q, and that what matters for the argument is assumption (*), the *truth* or *falsity* of P, Q,... (rather than, say, their *meanings*), and the *logical connectives* involved (¬ (not), ∨ (or), & (and), ⊃ (implies), and ~ (if and only if)). To convert the ideas behind these two arguments into a *decision procedure* for propositional logic, we first display the *truth tables* for the logical connectives (writing T, F for the *truth-values True* and *False*).

The truth tables

(¬: *negation*)   ¬P   This tabulates the fact that:

F T   if P is true, ¬P is false;
T F   if P is false, ¬P is true.

(∨: *disjunction*) P ∨ Q   This tabulates the fact that:

T T T   if P, Q are both true, P ∨ Q is true;
T T F
F T T   if at least one of P, Q is true, P ∨ Q is true;
F F F   if P, Q are both false, P ∨ Q is false.

(&: *conjunction*) P & Q   This tabulates the fact that:

T T T   if both P and Q are true, P & Q is true;
T F F
F F T   otherwise P & Q is false.
F F F

(⊃: *implication*) P ⊃ Q   This tabulates the fact that:

T T T
T F F   if P is true and Q false, P ⊃ Q is false;
F T T   otherwise P ⊃ Q is true.
F T F

(~: *equivalence*) P ~ Q   This tabulates the fact that:

T T T
T F F   if both P and Q have the same truth-value, P ~ Q is
F F T   true; otherwise P ~ Q is false.
F T F

These truth tables capture accurately the meanings attached to the logical connectives in classical mathematics: if the reader doubts this, reflection on some examples of mathematical reasoning should soon convince him. (See, for example, Rosser (1953), Chapter 2, especially §4.)

We can use these basic truth tables to construct truth tables for more complex propositions: thus, consider (P ⊃ (Q & ¬P)) ⊃ Q.

	P	Q	¬P	Q & ¬P	P ⊃ (Q & ¬P)	(P ⊃ (Q & ¬P)) ⊃ Q
(i)	T	T	F	F	F	T
(ii)	T	F	F	F	F	T
(iii)	F	T	T	T	T	T
(iv)	F	F	T	F	T	F
	(1) (2)		(3)	(4)	(5)	(6)

To consider all possible combinations of truth values we need, in this example, *four* lines, (i)–(iv). Column (3) is computed from column (1) by the truth table for ¬; column (4) comes from (2) and (3) by the table for &; column (5) comes from (1) and (4), by ⊃; and (6) comes from (5) and (2) by ⊃. The order in which the steps are conducted is determined by the number of occurrences of proposition letters (P, Q, etc.) in the *scope* of the logical connective. The *scope* of a connective is indicated by brackets where there might be any ambiguity (e.g. P ⊃ Q ⊃ R might be (P ⊃ Q) ⊃ R or P ⊃ (Q ⊃ R)). In the above example the scope of ¬ is P; the scope of & is Q and ¬P; the scope of the first ⊃ is P and (Q & ¬P); the scope of the second ⊃ is P ⊃ (Q & ¬P) and the last Q. The order in which the steps are taken in constructing a truth table is determined by the scopes of the connectives occurring in the formula; we deal first with the connective(s) with smallest scope, and so on up to the connective whose scope is the whole formula — called the *main* connective.

The full truth table, in our example, then can be made to look like this (with the steps numbered as before):

	(P	⊃		(Q	&	¬	P))		⊃		Q
(i)	T	F		T	F	F	T		T		T
(ii)	T	T		F	F	F	T		T		F
(iii)	F	T		T	T	T	F		T		T
(iv)	F	T		F	F	T	F		Ⓕ		F
	(1)	(5)		(2)	(4)	(3)	(1)		(6)		(2)

It will be noticed that under the main connective, line (iv) has the truth-value F, False. If under the main connective *all* lines have the value T(True) this means that there is no possible assignment of propositions/truth-values to P, Q which will make the whole formula false (so 'in all possible worlds' it is true). In this case it is a sound logical principle and we call it a *tautology*. If under the main connective *all* lines have the value F(False) we call it a *contradiction* (it cannot be true in any possible world). And if under the main connective some lines have the value T and some F the formula is *contingent* (true in some possible worlds, false in others). So our example is contingent. Only tautologies are sound logical principles.

EXERCISE 3.2.   Construct the full truth tables for the following principles and decide whether they are tautologies:

(i)   (P ⊃ (Q & ¬P)) ⊃ (P ⊃ Q).

(ii)   (P ⊃ (Q & ¬P)) ⊃ (Q ⊃ P). ∎

A short truth table method

It is clear that we can construct the full truth table for any formula of propositional logic and then simply inspect the column of truth values under the main connective to decide whether it is a tautology. We can often save ourselves the labour of writing out a full truth table by using some short variation: we now explain one of these.

Consider the formula ((A ⊃ B) ⊃ A) ⊃ A. If it is *not* a tautology there is at least one F in the column under its main connective. Then by the truth table for ⊃, the formula (A ⊃ B) ⊃ A must be T whilst A is F. But if A is F and (A ⊃ B) ⊃ A is T, then A ⊃ B must be F. But if A is F, then A ⊃ B cannot be F. In summary, numbering the truth values in the order in which they were determined:

	((A	⊃	B)	⊃	A)	⊃	A
	F	F		T	F	F	F
	(3)	(4)		(2)	(3)	(1)	(2)

The assumption that there is at least one F under the main connective of the given formula leads to a contradiction; hence the formula is a tautology.

Consider another example:

(P	⊃	(Q	&	¬	P))	⊃	(Q	⊃	P)
F	T	T	T	T	F	F	T	F	F
(4)	(2)	(4)	(6)	(5)	(4)	(1)	(3)	(2)	(3)

If we assume an F under the main connective all the other truth values come out as detailed; clearly it is *possible* for the whole formula to have F under its main connective, so it is not a tautology.

This method often works, and saves considerable labour. In general terms it is this: assume the formula has an F under its main connective; on this assumption, using the relevant truth tables, work out the truth values of its component sub-formulas dealing first with the connectives of greatest scope and lastly with those of least scope; if the resulting combination of truth values is possible, the formula is *not* a tautology; if it is impossible, the formula *is* a tautology. The idea is this: a logical principle is sound provided there is no possible world (possible interpretation of its proposition letters) in which it is false — provided it has *no counter-example.*

EXERCISE 3.3.   Check by the shortest method applicable the soundness of the following formulas and say whether your answer conflicts with your mathematical intuition.

(i)   (P ⊃ Q) ⊃ ((Q ⊃ R) ⊃ (P ⊃ R)).

(ii)   (P ⊃ Q) ⊃ (¬Q ⊃ ¬P); (¬Q ⊃ ¬P) ⊃ (P ⊃ Q).   (cf. Rosser pp. 33, 36)

(iii) (P ⊃ Q) ⊃ ((¬P ⊃ Q) ⊃ Q).

(iv)   (P ⊃ Q) ⊃ ((P ⊃ ¬Q) ⊃ ¬P).   (cf. Rosser p. 36)

(v)   (P ⊃ R) ⊃ ((Q ⊃ R) ⊃ ((P ∨ Q) ⊃ R)).   (cf. Rosser p. 34)

(vi)   (¬P ⊃ P) ⊃ P.

(vii) (P & (P ⊃ Q)) ⊃ Q.

(viii) P ⊃ (Q ⊃ (P & Q)).

(ix)   (P ⊃ Q) ⊃ (Q ⊃ P).

(x)   ((P ⊃ Q) ⊃ (Q ⊃ R)) ⊃ (P ⊃ R).

(xi)   ¬(P ⊃ Q) ⊃ (¬P ⊃ Q).

(xii) ((P ⊃ Q) ⊃ (R ⊃ S)) ⊃ ((¬S ⊃ Q) ⊃ (P ⊃ ¬R)). ∎

EXERCISE 3.4.
  (i)   Find illustrations of three of the sound principles above in the mathe-
        matics you already know.
  (ii) Find an illustration of a true theorem $P \supset Q$ whose converse $Q \supset P$ is
        false.
(Note: Rosser (1953) gives numerous illustrations in his Chapter II). ■

Some useful equivalences

We conclude these remarks on propositional logic with exercises proving some
useful equivalences and a result which these yield.

EXERCISE 3.5.   Show by truth tables that the following *equivalences* are
tautologies:

  (i) $(\neg\neg P) \sim P$;

  (ii) $(\neg P \vee Q) \sim (P \supset Q)$;

  (iii) $(P \vee Q) \sim (\neg P \supset Q)$;

  (iv) $(P \,\&\, Q) \sim \neg(\neg P \vee \neg Q)$;

  (v) $(P \vee Q) \sim \neg(\neg P \,\&\, \neg Q)$;

  (vi) $(P \supset Q) \sim \neg(P \,\&\, \neg Q)$;

  (vii) $(P \sim Q) \sim ((P \supset Q) \,\&\, (Q \supset P))$. ■

EXERCISE 3.6.   Using the equivalences in Exercise 3.5 show that any principle
of propositional logic is equivalent to one containing *only* the connectives $\supset$ and
$\neg$. ■

Predicate logic

We now consider briefly the scope for extending these methods to predicate
logic. For a principle of predicate logic, say $R(a, a) \supset \exists b R(b, a)$, to be *valid*, it
must be *true* for *any* (non-empty) class of objects and any relation R. Thus
deciding whether such a principle is valid necessitates considering by some means
all classes of objects (call them *domains*) and all relations defined over such
domains.

   We may think of $\exists x A(x)$ as an extended disjunction

$$A(s_1) \vee A(s_2) \vee \dots \vee A(s_i) \vee \dots;$$

its length depending on the number of elements (s_i) in the domain being considered. Similarly, $\forall x A(x)$ may be thought of as an extended conjunction

$$A(s_1) \,\&\, A(s_2) \,\&\, \dots \,\&\, A(s_i) \,\&\, \dots .$$

If the domain we are considering is finite, with n members, we can thus translate any formula of predicate logic into one of propositional logic: we simply replace ∀ and ∃ by n-place conjunctions and disjunctions respectively, replacing the quantifier of least scope first.

Working this out for domain D containing only two members 1, 2, we translate $\exists b \forall a R(a, b)$ as follows:

(1st step) $\exists b(R(1, b) \,\&\, R(2, b))$;

(2nd step) $[R(1, 1) \,\&\, R(2, 1)] \lor [R(1, 2) \,\&\, R(2, 2)]$.

The resulting formula is easily evaluated by truth tables, and it is easy to assign truth values which make it false.

EXERCISE 3.7.   For the domain D given above, translate

$$\forall a \; \exists b \; R(a, b) \supset \exists b \; \forall a \; R(a, b)$$

into a formula of propositional logic and test by truth tables whether the resulting formula is true in D. ■

Although this procedure is unwieldy (and rapidly becomes more so as D increases in size!), it is possible in principle to decide for any fixed-size finite domain D and any formula of predicate logic F, whether F is true in D. But clearly we *cannot* even in principle consider separately *all* finite domains (since there are infinitely many of these), nor can we, for an *infinite* domain D, translate an arbitrary formula of predicate logic into one of propositional logic (since it would be of infinite length). The method, then, is useful for counter examples but not as a test of general validity.

For an arbitrary formula F to be *logically valid* it must be true in *all* domains, finite and infinite, but it is invalid if there is *one* domain in which it is false. If we set out to check domains of increasing size, starting with the domain of one member, we may find one in which it is false, then we know it is invalid. But we can never complete this search, so it will never establish validity. This requires other methods — where possible. So the scope for extending truth-table methods to predicate logic are very limited: once we move from the *finite* to the *infinite*, as we shall see in the next chapter, we move into a totally different realm.

4
PROBLEMS IN THE FOUNDATIONS OF MATHEMATICS

The purpose of this chapter is to sketch the historical background which generated the problems, the novel methods, and the surprising results expounded in the remainder of this work. The last portion of the chapter is an introduction to the concepts of formalization and of a formal metamathematics.

Intuitions and axioms

We have already seen in Chapters 1 and 2 how we proceed to axiomatize a given body of knowledge (or theory): we organize our beliefs in this domain so that all of them can be deduced from certain fundamental propositions which are *true* – or seem to be – and these we take as our axioms. Provided we choose our axioms well (i.e. they really are *true*) and we use *reliable* principles of inference, then we can be confident that the *theorems* deduced from those axioms are *true*. What guides us in this process is our *intuitions*, our intuitions about numbers, or about physical space (Euclid's *Elements* has been a model of the axiomatic deductive method since the third century BC and certainly Euclid believed that his axioms were true of physical space), or about consumer preferences (cf. Arrow 1951), or whatever the subject is. However, we sometimes deduce a theorem which causes us to revise either our axioms, or our rules of inference, or our intuitions; in the interplay between these three there is no rule for deciding which has priority. Intuition is our starting point, but it is not always a reliable guide. Dramatic demonstrations of this fact arose from Cantor's theory of infinite sets (1895-7) and from Frege's attempt to deduce mathematics from logic (1884) and (1893). We briefly explain both of these now.

Infinite sets and the diagonal method: Cantor's set theory

(Note that we shall use the word 'set' to mean simply a collection or class of objects, e.g. the set of men or the set of positive integers. If an object x *belongs to* a set S we shall write $x \in S$. If x does *not* belong to S we write $x \notin S$ or $\neg x \in S$. We give here no more of Cantor's set theory than is essential for our limited purposes.)

Two children, Albert and Bertrand, who cannot yet count, dispute which has the larger number of tin soldiers. How can we settle the matter for them — without counting of course? We can do so by returning soldiers *in pairs* to the toy box, one from each army, until one child has no soldiers left. Whoever has soldiers remaining had the largest army. If both armies disappear simultaneously they each contained the same number of soldiers. By our action (*pairing* the soldiers) we establish what is called a *one-to-one mapping* or a *one–one mapping* or a *1–1 mapping* between the two armies. If, for every member of A's army (call it set A) we can find a (unique) partner from B's army (call it set B) we say this mapping is from A into B. Conversely for B into A. If, by this 1–1 mapping (from A into B) *every* member of B is associated with a (unique) member of A, it is called a 1–1 *onto* map or a *1–1 correspondence*: if we can find a 1–1 correspondence between two sets then they have the *same number* of members.

EXERCISE 4.1.   Is there a one–one correspondence between;

   (i)   the set of human males and the set of sons;

  (ii)   the set of hearts and the set of livers in live animals;

 (iii)   the set of positive integers and the set of negative integers? ∎

It is the generalization of these ideas to *infinite* sets, a generalization which seems very natural once the basis of counting finite sets is understood, which gives rise to Cantor's set theory — or as some have called it 'Cantor's Paradise'. For centuries it had been realized that infinite sets differed from finite sets in that they could be put into 1–1 correspondence with a *proper* part (i.e. not the whole) of themselves, e.g. the positive integers and the even positive integers:

$$
\begin{array}{cccccc}
1 & 2 & 3 & 4 & 5 & \ldots \\
| & | & | & | & | \\
2 & 4 & 6 & 8 & 10 & \ldots
\end{array}
$$

Or again, there was 'Galileo's paradox' that the positive integers could be put into 1–1 correspondence with their squares:

$$
\begin{array}{ccccccc}
1 & 2 & 3 & 4 & 5 & 6 & \ldots \\
| & | & | & | & | & | \\
1 & 4 & 9 & 16 & 25 & 36 & \ldots
\end{array}
$$

Any set which can be put into 1–1 correspondence with the natural numbers 0, 1, 2, 3, 4,... we shall call *enumerable, denumerable,* or *(d)enumerably infinite*, and any particular such correspondence we shall call an *enumeration*.

EXERCISE 4.2.   Show that there is a 1–1 correspondence between the positive integers and each of the following sets:

(i)   the prime numbers;

(ii) the pairs of positive integers;

(iii) the finite sets of positive integers;

(iv) the positive rational numbers. ∎

It might be thought that all infinite sets have the 'same size', i.e. can be put into 1-1 correspondence with the natural numbers and are therefore denumerable, but by Cantor's *diagonal method* we can show this is not so. Consider the real numbers x in the interval $0 < x < 1$ and suppose they are each expressed by a (denumerably) infinite decimal expansion (so $\frac{1}{2}$ is *not* 0.5 but is 0.4999...; $\frac{1}{4}$ is not 0.25 but is 0.2499...). Now, let us assume *there is* a 1-1 correspondence between these real numbers and the natural numbers, given in the following enumeration (each x_{ij} is a digit 0, ..., 9)

$$1 - 0.x_{11}x_{12}x_{13}x_{14} \ldots$$
$$2 - 0.x_{21}x_{22}x_{23}x_{24} \ldots$$
$$3 - 0.x_{31}x_{32}x_{33}x_{34} \ldots .$$

Now consider the 'diagonal' real number obtained from this supposed enumeration by changing each x_{ii} to x'_{ii} (where x'_{ii} is $x_{ii} + 1$ unless $x_{ii} = 9$ when $x'_{ii} = 0$): the real number

$$0.x'_{11} x'_{22} x'_{33} \ldots$$

differs (in at least one place in its decimal expansion) from every other real number in the enumeration. So the assumption that the real numbers x in the interval $0 < x < 1$ can be put into 1-1 correspondence with the natural numbers has led to a contradiction: so these real numbers are not *denumerable*. We shall call sets which can be put into 1-1 correspondence with these reals *nondenumerable* or *non-denumerably infinite*. Once this diagonal construction is accepted you are admitted to Cantor's paradise — an ever ascending hierarchy of infinities. For example, by the same construction it can be shown that the real functions (i.e. functions from real numbers into real numbers) comprise an even larger infinite set than the real numbers themselves, and so on. ∎

EXERCISE 4.3.   Show that there is no 1-1 correspondence between the real numbers and the one-place real functions. (*Hint*: assume *there is* a 1-1 correspondence and let the real number x be mapped to the one-place real function $f_x(y)$. A one-place function is a function with exactly *one* argument.) ∎

Some people, especially the intuitionists, believe that this construction can quickly lead one astray. Most mathematicians have found it enormously fruitful and we shall continue to use it, but we note here that it is *not* included among the *finitary* methods to which we shall shortly refer.

Explicit logical axioms for mathematics: Frege's logicism

Frege held that the truth of mathematics could be deduced from the principles of logic – that mathematical notions, like number and function, could be defined in terms of logical notions, like concept and relation, and that this having been done, the truths of mathematics could be deduced from logical principles *alone*. The view that the fundamental propositions from which mathematics can be deduced are the propositions of logic is known as *logicism*. The logicist programme was carried out independently by Russell and Whitehead in their *Principia Mathematica* (1910–13), but for our purposes Frege is the more important source, especially his *Die Grundgesetze der Arithmetik* (1893, 1903).

In order to establish his thesis Frege proceeded in the manner of our Chapters 1 and 2 to 'dig out' the mathematical and logical 'foundations' of classical mathematics and with the aid of appropriate definitions – including using the notion of 1–1 correspondence to define number – he deduced a large part of mathematics from what seemed secure logical foundations by means of reliable logical principles. Although he expressed some misgivings about one of his axioms, his axiom V, he was generally confident of his foundations. He was, however, much more meticulous than we were in Chapters 1 and 2. He realized not only that ordinary language is hopelessly cumbersome and imprecise for doing mathematics, but also that existing mathematical symbolism was ambiguous and confused. To remedy this he invented a new, meticulously defined symbolism in which to express mathematical propositions: (a) he specified exactly which symbols belonged to his 'language' (for expressing mathematical propositions); (b) he specified exactly which combinations of symbols were 'meaningful' and which were nonsense (by rules like ordinary grammatical rules); and (c) he said in his symbolism (which displayed accurately the logical structure of the mathematical proposition it expressed) exactly what were his axioms and his rules of inference.

He was also careful to give *full* proofs, i.e. proofs in which *everything* was explicit; if a rule of inference was used its use was explicitly stated; if a logical axiom was used its use too was explicit. In constructing a proof one could not take for granted some principle which everyone accepted as true or appeal to some aspect of the meaning of a symbol, unless these were explicitly deducible from the axioms. For example, one could not assume for some mathematical proposition P, that 'P or not P' (the Law of excluded middle) unless this was explicitly deducible from the axioms: one could not assume that '(P and Q) implies P' unless this was explicitly deducible from the axioms. Such a notion of proof is very different from what mathematicians are accustomed to calling a proof. In this respect a *proof* for Frege is a sequence of formulas (of his symbolism) such that each formula is either an axiom or is explicitly deducible from preceding members of the sequence by one of the rules of inference. It should be clear from this that although Frege regarded his axioms as *true* and his rules as *sound* principles of inference, it is possible to check whether a sequence of formulas is a proof by ignoring their intended meaning and by attending to them simply as formal objects. We shall exploit this possibility shortly.

Russell's paradox discovered

One advantage of making everything explicit is that if something goes wrong you can immediately trace the source. Hardly had Frege's *Grundgesetze* been finished when Bertrand Russell discovered that something was indeed wrong, that Frege's system entailed a contradiction. This result, now generally known as Russell's paradox, was first published in Frege (1903). Russell's paradox arises as follows:

Some sets are members of themselves (e.g. the set of infinite sets) some are not (e.g. the set of men). Now consider the set of *sets which are not members of themselves*, call it T. Either $T \in T$ or $T \notin T$. Suppose $T \in T$, then T is a set which is not a member of itself, so $T \notin T$. By *reductio ad absurdum* we have proved that $T \notin T$. But now if $T \notin T$, T is a set which is not a member of itself, so $T \in T$! Contradiction.

EXERCISE 4.4.   Derive a similar contradiction from the assumption that there is a village in which there is a barber (male) who shaves all and only those men who do not shave themselves. ■

This contradiction can be deduced in Frege's system and Frege traced it to his axiom V, which says, roughly:

Two concepts determine the same set if and only if they apply to the same objects. E.g. the concepts 'equilateral triangle' and 'equiangular triangle' determine the same set: the concepts 'positive integer' and 'rational number' do not determine the same set.

A theory in which a contradiction can be proved is one in which you can prove *anything*! Remember, $((B \ \& \ \neg B) \supset A)$ is a tautology.) But however we try to escape this contradition, whether by rejecting the law of excluded middle for sets or by denying that a set can be a member of itself or by abandoning Frege's axiom V, intuitions basic to classical mathematics have to be rejected. Russell's paradox is not a superficial matter for classical mathematics but goes very deep. Numerous related 'paradoxes' were discovered at about the same time including Cantor's in 1899, Burali-Forti's in 1897, Richard's in 1905, Berry's in 1906 and Grelling's in 1908. (These are all given in Mendelson 1964, p. 2 f.) Some of these derived similarly deep-seated contradictions in Cantor's set theory. As Gödel puts it in (1944) the paradoxes reveal 'the amazing fact that our logical intuitions (i.e. intuitions concerning such notions as: truth, concept, being, class, etc.) are self-contradictory.'

Finitism and formalism: Hilbert's programme

Faced with such serious flaws mathematics needed radical treatment. It was in response to this need that *Hilbert's programme* was formulated. Frege had already insisted on *explicit* symbolism, *explicit* axioms and inference rules, and *explicit* proofs. Hilbert went further; he proposed to treat such an explicit theory as itself an *object* of mathematical study and, using suitable method, to decide whether it was *consistent*. This approach is generally known as *metamathematics* or *proof theory*. His proposal contained two elements.

The first was to ignore any intended meanings or possible interpretations of the symbols, formulas, or proofs and to consider them purely as *formal objects*, investigating their *structural* properties in order to show that there is no formula A such that both A and $\neg$A are provable. This gives us the idea of a *formal theory*: a set of symbols, of rules for constructing formulas, and of rules for constructing proofs, all devoid of meaning or content. This is part of Hilbert's *formalism*.

The second was to ensure that you can always check in a *finite* number of 'mechanical' steps whether a sequence of formulas is a *proof* in the formal theory and, in applying mathematical methods to such a formal theory, to avoid all those inferences involving the infinite and which have led to trouble. This excluded especially the methods peculiar to Cantor's set theory and was part of Hilbert's *finitism*. The idea was to use only 'safe' methods since there is little point in proving mathematics – or part of it – consistent by methods which are themselves unreliable. He called the methods which are appropriate to meta-mathematics *finitistic* or *finitary*. It is difficult to say exactly what these include, but it should not mislead the reader at this stage to say that in effect they are the methods of elementary arithmetic including for example mathematical induction. Subsequently, mathematicians have expanded the methods they count as 'safe' and have thus generalized metamathematics; it is still a field of active research.

It is clear that contradictions must be eliminated from mathematics, and Hilbert's programme deserves to be taken seriously. In the next section we give a definition of 'formal theory' in Hilbert's terms. Having done this we shall shortly specify a formal theory, called **N** for that fragment of mathematics comprising elementary number theory and we shall then apply Hilbert's recommended methods to **N**. The results are not at all what he expected!

Decidable predicates and formal theories

It is easy to check in a finite number of mechanical steps if an arbitrary positive integer n is a *prime number*, or *is even*, or *is a perfect square*. There is a 'mechanical routine' for answering these questions; no ingenuity, inventiveness, or mathematical judgement is required; indeed a computer could be programmed to decide them. So these are called *decidable* predicates or properties (sometimes *effective* or *constructive* predicates). There is no such mechanical routine for deciding whether, for arbitrary n, a sequence of n successive 7s occurs in the decimal expansion of π, or whethere an arbitrary mathematical proposition is true, or whether an arbitrary (informal) proof is sound.

Now, if it is to be possible to check in a finite number of mechanical steps whether a sequence of formulas is a proof in a formal theory — if this is to be *decidable* — clearly, it must also be decidable what counts as a symbol, a formula, an axiom, and a rule of inference: it must also be decidable whether the ith formula follows by one of the rules from some preceding formulas in the sequence. These are the requirements which determine our definition of *formal theory* (or, *axiomatic formal theory*, as we shall sometimes say).

DEFINITION 4.1. A *formal theory* **T** consists of a set of symbols, formulas, axioms, and rules of inference:

Symbols. A finite or denumerable set of symbols must be specified as the symbols of **T**. It must be decidable what is a symbol of **T**. A finite sequence of its symbols is called an *expression* of **T**.

Formulas. Which expressions are formulas must be specified. It must be decidable which are formulas. Formulas are usually specified by means of 'formation rules'.

Axioms. Which formulas are axioms must be specified. It must be decidable which formulas are axioms. (Note: there may be denumerably many axioms.)

Rules of inference. A *finite* set of rules of inference $R_1, \ldots, R_n$ must be specified. (In a formal theory **T**, rules of inference are simply relations between formulas.) For each R_i there must be a unique positive integer p such that for every set of p formulas and each formula ϕ it is decidable whether ϕ stands in the relation R_i to the given p formulas: if it does ϕ is called an *immediate consequence* of the p formulas by R_i in **T**. ■

An example

A simple example of a formal theory is the following theory, called **E**.

Symbols

(i) P_i is a symbol of **E**, for all $i \geqslant 0$;

(ii) $\supset, \neg, (,)$ are symbols;

(iii)   there are no other symbols.

Formulas

   (i)   Each P_i is a formula;

  (ii)   if A and B are formulas, so are $\neg$A and (A $\supset$ B);

 (iii)   there are no other formulas.

Axioms. If A, B, and C are formulas the following are axioms:

   (i)   (A $\supset$ (B $\supset$ A));

  (ii)   ((A $\supset$ (B $\supset$ C)) $\supset$ ((A $\supset$ B) $\supset$ (A $\supset$ C)));

 (iii)   (($\neg$B $\supset$ $\neg$A) $\supset$ (($\neg$B $\supset$ A) $\supset$ B));

 (iv)   there are no other axioms.

Rules of inference. There is only one rule of inference. If A and B are formulas, B is an immediate consequence of (A $\supset$ B) and A.

If we interpret each P_i as a proposition letter, '$\supset$' as 'implies', and '$\neg$' as 'not', then the formulas belong to propositional logic, the axioms express tautologies and the rule is *modus ponens*. Although it is by no means obvious, any tautology is provable in **E**. (Remember, $\vee$, &, and $\sim$ are definable in terms of $\neg$ and $\supset$ alone, by Exercise 3.6.) In fact, the only thing **E** is good for is the proving of tautologies.

The notion of formal proof and first order theory

We now give some definitions relating to formal theories. They are collected together for ease of reference and nearly all the exercises relating to them occur in the course of subsequent chapters. At this stage the reader should attempt to see the appropriateness of the definitions in the light of the *formalist* and *finitist* elements in *proof theory*.

DEFINITION 4.2.   A *formal proof* in **T** is a *finite* sequence of formulas such that each member of the sequence is either an *axiom* of **T** or an *immediate consequence* of preceding formulas. A *theorem* of **T** is the last member of such a sequence. 'B is a theorem of **T**' is written $\vdash_{\mathbf{T}}$B. (If the context shows clearly which formal theory is intended the subscript **T** is often omitted.) ∎

DEFINITION 4.3.   A formal theory **T** is *consistent* provided there is no formula B such that both $\vdash_{\mathbf{T}}$B and $\vdash_{\mathbf{T}}$ $\neg$ B. If it is *decidable* for an arbitrary formula B, of **T**, whether $\vdash_{\mathbf{T}}$B, the formal theory **T** is said to be *decidable*; otherwise *undecidable*. ∎

DEFINITION 4.4.   A formal theory **T**′ *having the same symbols* as a formal theory **T** is called an *extension* of **T** if every theorem of **T** is a theorem of **T**′; **T** is then called a *subtheory* of **T**′. ∎

EXERCISE 4.5. For the formal theory E and the suggested interpretation show that:

(i) the axioms are tautologies;

(ii) if A and (A ⊃ B) are tautologies then B is a tautology: hence,

(iii) if B is a theorem of E, B is a tautology; hence,

(iv) E is consistent;

(v) under the assumption that if A is a tautology, then A is a theorem of E, the theory E is decidable. ∎

Throughout the remainder of this work we shall consider only *first-order* formal theories which have quantifiers over individual variables, but are lacking quantifiers over predicates, or sets, or functions. In effect, this means formal theories with the propositional and predicate logic of the theory N given explicitly below. Since all the formal theories we consider are first-order and all (but one) are *axiomatic* (i.e. it is *decidable* what counts as an axiom), we shall normally omit both terms and speak simply of *formal theories* (and even, when the context leaves no room for ambiguity, simply *theories*).

Gödel numbering

Having outlined the elements of Hilbert's programme, both its *formalism* and its *finitism*, we shall shortly present a formal theory for arithmetic and proceed to study it by *finitary methods*. Remember, finitary methods are in effect those belonging to arithmetic. We shall therefore be using the methods of arithmetic to *study* arithmetic − through its formal theory. This process may seem circular and self-defeating, unless the reader is careful to maintain a clear distinction between several levels; as Kleene puts it (1952, p. 65):

In the full picture, there will be three separate and distinct 'theories': (a) the informal theory of which the formal theory constitutes a formalisation; (b) the formal system or object theory; and (c) the metatheory, in which the formal system is described and studied.

There is no alternative here to keeping a clear head − always being careful about which realm we are working in: this is all the more important when it is realized that the methods of the metatheory (c) belong to the informal theory (a) and may therefore be, in some sense, expressible in (b). Indeed this 'self-reference' is the clue to Gödel's discoveries and the key tool is his invention now generally known as *Gödel numbering*.

DEFINITION 4.5. Suppose we have specified the symbols of some formal theory (we use N of our next chapter as our example). Then to each of its symbols s_i, we assign a *unique* number, called its *gödel number*, thus:

$\supset$	&	$\vee$	$\neg$	$\forall$	$\exists$	$=$	$+$	$\cdot$	$'$	0	(	)	a	b	c	...
3	5	7	9	11	13	15	17	19	21	23	25	27	29	31	33	...

We assume the denumerable number of variables a, b, c, ... are given in some uniform way so that it is decidable what counts as a variable. The even numbers are left unused in case we would wish to add more symbols to the theory. We write the gödel number of s_i as $gn(s_i)$. To an *expression* $s_0 s_1 \ldots s_n$ (and hence to a *formula*) we assign the gödel number as follows:

$$gn(s_0 s_1 \ldots s_n) = 2^{gn(s_0)} \cdot 3^{gn(s_1)} \cdot \ldots \cdot p_n^{gn(s_n)}$$

where p_i is the ith prime, $p_0 = 2$. ∎

For example $gn(\neg 0 = 0) = 2^9 \cdot 3^{23} \cdot 5^{15} \cdot 7^{23}$. Note that gödel numbers are always positive, and gödel numbers of expressions are always even.

EXERCISE 4.6.   Under this assignment of gödel numbers:

(i)   give the *gn*s of the following expressions:

(a) $\neg a \forall$,   (b) $\forall a (a = 0)$;

(ii)   could two different expressions get the same *gn*?

(iii)   could the same positive integer be the *gn* of both a symbol and an expression?

(iv)   what is the *gn* of $\supset$

(a) as a symbol?   (b) as an expression? ∎

DEFINITION 4.6.   To a *finite sequence of expressions*, $e_0, e_1, \ldots, e_m$ (and, hence, to a *formal proof*), we assign the gödel number:

$$gn(e_0, e_1, \ldots, e_m) = 2^{gn(e_0)} \cdot 3^{gn(e_1)} \cdot \ldots \cdot p_m^{gn(e_m)}. ∎$$

EXERCISE 4.7.   Under this assignment of gödel numbers,

(i)   could two different sequences of expressions get the same *gn*?

(ii)   could the same positive integer be the *gn* of both a sequence of expressions and either a symbol or an expression?

(iii)   is every positive integer a *gn*? ∎

This gödel numbering provides a 1–1 mapping from the set of symbols, expressions, and sequences of expressions of **N** *into* the positive integers. There are of course an infinite number of other possible ways of gödel-numbering **N**, but we need only consider the one we have described. Note that by the uniqueness of prime factorization, given any positive integer k, it is decidable whether k is the *gn* of a symbol, expression or sequence of expressions, and if so which.

EXERCISE 4.8.   Under our assignment of gödel numbers of what, if anything, are the following gödel numbers:

(i) 864;   (ii) 865;   (iii) 866? ■

Arithmetization of metamathematics

Using this gödel numbering it is possible to translate metamathematical properties and relations into arithmetic ones. This is called the *arithmetization of meta-mathematics* and is the purpose for which Gödel invented the procedure. To give a trivial example, let x be any symbol of **N**, then 'x is a variable' is true if and only if '2 $\nmid$ *gn*(x) and *gn*(x) $\geqslant$ 29'.

More interesting examples get rapidly more complicated, and although it is very complicated, it is possible to find an expression of (informal) arithmetic, call it Proof (x, y), which is true exactly when 'x is the gödel number of a proof of the formula with gödel number y'. This informal expression will be formal-izable in **N**, so it is not hard to believe that with some ingenuity it will be possible to find formulas of formal arithmetic **N**, which under the 'usual inter-pretation' of their symbols express something which may or may not be true of themselves and this opens the way for constructions like Russell's paradox. This is precisely the possibility Gödel exploits to such effect in deriving his famous results.

But we must not leap ahead too fast. The motivation for what we are doing should now be clear, especially why the sometimes cumbersome detail of the next two chapters is necessary. Formalizing a theory enables it to be studied by means of a whole new set of tools and often yields startling new results, so we now present the theory **N** explicitly.

5

THE FORMAL THEORY N INTRODUCED

The axiomatization of elementary number theory which we now present is chosen because it is reasonably easy to handle and well-worked. Except for minor variations the formalization is that used by S. C. Kleene and is to be found in his *Introduction to Mathematics* (1952) and his *Mathematical Logic* (1967). We first present the symbols, formation rules, axioms, and rules of inference of **N** and then show that familiar logical moves are possible in **N** by proving several derived rules of inference; in particular the Deduction Theorem is discussed in detail.

The symbols of the theory

(The list of symbols we present could be made shorter, but the economy is not worth the effort. The list could also be made longer, but we take the view that extra symbols ought when possible to be introduced by *definition* (or abbreviation), a process we discuss later.)

DEFINITION 5.1. The *formal symbols* of the theory **N** are given below with their intended interpretation in parentheses:

$\supset$ (implies); & (and); $\vee$ (or); $\neg$ (not); $\forall$ (for all); $\exists$ (there exists);
$=$ (equals); $+$ (plus); $\cdot$ (times); $'$ (successor); 0 (zero);

additionally we also use as *symbols*:

a, b, c, ... (variables); (,) (parentheses). ∎

The formal symbols are chosen to agree with convention and with an intended interpretation in mind (i.e. elementary number theory), but of course *this interpretation is no part of the formal theory*. The variables are interpreted as ranging over the natural numbers and we suppose we have an infinite list of variables available.

EXERCISE 5.1. Do the following *symbols* belong to our formal theory?

(i) $<$ (ii) $\sim$ (iii) $|$ (iv) 2 (v) $=$ (vi) $\neq$

(vii) {   (viii) ∈   (ix) →   (x) ! ∎

DEFINITION 5.2.   A finite sequence of (occurrences of) formal symbols of **N**
is called a *formal expression* of **N**. The empty sequence will *not* count as a
formal expression. ∎

EXERCISE 5.2.   Are the following *expressions* of our formal theory?

  (i) (a + b)   (ii) a + b   (iii) (a + (b   (iv) 0

  (v) ∀   (vi) The empty sequence of symbols

  (vii) 0 ∀ 0 0 =   (viii) An finite sequence of 0s

  (ix) Pr(2)   (x) All the variables listed in succession. ∎

Formation rules

According to Definition 4.1, the next step is to specify the *formulas* of the formal
theory. In our approach to arithmetic this has to be done in two stages, because
the theory **N** naturally involves various 'algebraic' expressions in making equations; these special expressions are called *terms* in general.

DEFINITION 5.3.   The *terms* of the theory **N** are given by these rules:

  (1)   0 is a *term*;

  (2)   A variable is a *term*;

  (3)-(5)   If s and t are *terms*, then $(s + t)$, $(s \cdot t)$ and s' are *terms*;

  (6)   The only *terms* are those given by (1)-(5). ∎

By successive applications of these rules we can build up any term; thus, by
rule (1), 0 is a term; by (2), a is a term; by (5), $0'$ is a term; and by (3) $(0' + a)$
is a term. Terms can be of any (finite) length.

EXERCISE 5.3.   Which of the following are *terms* of our formal theory?

  (i) 0   (ii) (0)   (iii) $0'$ (iv) $(0' + 0)$

  (v) $0''$ (vi) 2   (vii) 2^2 (viii) $0''0'$

  (ix) $(a \cdot b)$ (x) $a \cdot b$ (xi) $((a \cdot b))$ (xii) (a = b)

  (xiii) rm(a, b)   (xiv) p_i (xv) ∃c(c = c) (xvi) 3!

  (xvii) (a!) ∎

DEFINITION 5.4.   The *formulas* of the theory **N** are given by these rules:

  (1)   if s and t are terms, then $(s = t)$ is a *formula*; a formula which contains
      no logical connective or quantifier is called a *prime formula*;

(2)-(5) if A and B are *formulas*, then (A ⊃ B), (A & B), (A ∨ B), and ¬A are *formulas*;

(6)-(7) if x is a variable and A is a *formula*, then ∀xA and ∃xA are *formulas*;

(8) the only *formulas* are those given by (1)-(7). ∎

So, clearly, the following is a sequence of formulas by virtue of this definition:

$$(a = b),$$
$$((c' + a) = b),$$
$$\neg(a = b),$$
$$\exists c((c' + a) = b),$$
$$(\exists c((c' + a) = b) \supset \neg(a = b)).$$

Note that in rules (6)-(7) above 'x' *stands for any formal variable of* N. It is functioning here as what is called a 'meta-variable'.

EXERCISE 5.4.   Which of the following are *formulas* of our formal theory?

(i) $0 = 0$ (ii) $(0 = 0)$ (iii) $(0 = 0')$ (iv) $a < b$

(v) Pr(a)   (vi) $c = \text{rm}(a, b)$ (vii) $A \supset B$ (viii) $P \supset Q$

(ix) $(A \supset B)$ (x) $(A \supset B)$ where A, B are formulas   (xi) $\forall xA$

(xii) $\forall a(0 = a)$ (xiii) $\forall a(0 = b)$ (xiv) $(0 = (0 = b))$

(xv) $\exists a \forall a(0 = b)$ (xvi) $\forall a \forall b \exists q \exists r(a = ((b \cdot q) + r))$

(xvii) $((a + 0) = a)$ (xviii) $(a)^0 = (0')$ (xix) $(a)^{b'} = ((a)^b \cdot a)$ ∎

This completes the specification of the formation rules for N; before we can give the *axioms* and *rules of inference* we need to explain the notions of *scope* of an *operator*, *free* and *bound* variables, and *substitution*.

Scope of an operator

The *operators* of N are,

$$\supset, \&, \vee, \neg, \forall x, \exists x, =, +, \cdot, ' \text{ (where x is any variable).}$$

Terms and formulas are built up using these operators in accordance with the rules given above. Note that

$$((0' + 0') = 0'')$$

contains four *occurrences* of the operator '.

DEFINITION 5.5.   For any term or formula, the *scope* of (an occurrence) of an operator is the expression or expressions to which the (occurrence of the) operator was applied in building up the term or formula by the rules of Definitions 5.3 and 5.4. ■

So the operators in $\exists c((c' + a) = b)$ have the scopes displayed by the heavy lines.

$$\exists c((c' + a) = b)$$

It should be clear that in any given term or formula the scopes of the operators can be recognized without ambiguity by pairing parentheses. However, formulas containing multiple parentheses are not always easy to read. To overcome this problem we shall often abbreviate terms or formulas by omitting parentheses so that they can be restored unambiguously by the following widely used convention: dealing first with all occurrences of the operator ′, then with all occurrences of ·, and so on in the order ′, ·, +, =, $\exists x$, $\forall x$, ¬, ∨, &, ⊃, restore parentheses to give any occurrence of an operator the *least* scope consistent with the resultant expression being a term or formula. We shall not always omit all parentheses in accordance with this convention if to leave some in improves readability. Our actual usage will parallel that of informal mathematics, for example

$$\exists c(a + c' = b) \supset \neg a = b$$

is probably the 'natural' abbreviation for

$$(\exists c((a + c') = b) \supset \neg(a = b))$$

by this convention.

EXERCISE 5.5.   Give the scope of the operators in the following formulas:

   (i) $((a + 0') = a)$;

  (ii) $\forall a \exists a(0 = b)$;

 (iii) $\forall a \forall b \exists q \exists r(a = b \cdot q + r)$;

 (iv) $\exists c(\exists c(c' + a = b) \supset \neg a = b + c)$;

  (v) $(A \supset B) \supset ((A \supset \neg B) \supset \neg A)$ where A, B are formulas. ■

This is not the only kind of abbreviation we shall use to make formulas easier to read. For $0', 0'', 0''',\ldots$ we shall write $1, 2, 3,\ldots$; for $\exists c(c' + a = b)$

we shall write a $<$ b; and there are others which will be explained as we come to them. It remains true that these abbreviations are *not* expressions of the formal theory but are shorthand for such expressions which we can always get back to if we wish.

Free and bound variables

DEFINITION 5.6.

    (i)   An *occurrence* of a variable x in a formula A (or term t) is said to be *bound* if that *occurrence* is in a quantifier $\forall$x or $\exists$x or is in the scope of a quantifier with the same variable; otherwise the *occurrence* is called *free*.

    (ii)   A bound occurrence of a variable is bound *by* the quantifier within whose scope it lies and which has the least scope, or, if it occurs in a quantifier, *by* that quantifier itself.

    (iii) If a variable x has a free occurrence in a formula A it is called a *free variable of* A and A is said to *contain* x *free*; similarly for bound variables. ■

Notice that the distinction between a free and a bound occurrence of a variable is always relative to the term or formula in which it is (at the moment) being considered as an occurrence. Thus in $\exists c(c' + a = b)$ the second occurrence of c is free when considered as an occurrence in c, c′, c′ + a, c′ + a = b by themselves, but bound as an occurrence in $\exists c(c' + a = b)$. Notice also that a variable x may be both a free and a bound variable of a formula A.

EXERCISE 5.6.   For each formula α, β answer questions (i)–(iv) below:

    α $\neg b = 0 \supset \exists q \exists r(a = b \cdot q + r \,\&\, \exists a(a' + r = b))$;

    β $\exists c(c' + 0' = a) \supset \neg(\exists c(a \cdot c = b) \,\&\, \exists c(a \cdot c = b'))$.

    (i)   Which occurrences of variables are free and which are bound?

    (ii) Considering only occurrences of the variable a, state in which terms and subformulas they are free and in which they are bound. (A *subformula* of a formula A is any (consecutive) expression in A which is itself a formula.)

    (iii) Is any variable both a free and a bound variable of the formula?

    (iv) By which quantifier is each bound variable bound?

  (Note. α expresses the existence of quotient q and remainder r, r $<$ b. β abbreviates to a $>$ 1 $\supset \neg(a|b \,\&\, a|b')$.) ■

Substitution

For what follows we need to define an operation of *substitution*. This should not be confused with the rule of substitution given on p. 57.

DEFINITION 5.7.   The *substitution of* a term t for a variable x *in* (or *throughout*) a term or formula A shall consist in replacing simultaneously each free occurrence of x in A by an occurrence of t. We shall express this by means of the following convenient metamathematical notation: writing 'A(x)' for the formula in which t is to be substituted for the variable x, we shall write the result of the substitution as 'A(t)'. ∎

Thus if x is a and A(x) or A(a) is

$$\exists c(c' + 0' = a) \supset \neg(\exists c(a \cdot c = b) \,\&\, \exists c(a \cdot c = b'))$$

then A(0) is

$$\exists c(c' + 0' = 0) \supset \neg(\exists c(0 \cdot c = b) \,\&\, \exists c(0 \cdot c = b'))$$

and A(c) is

$$\exists c(c' + 0' = c) \supset \neg(\exists c(c \cdot c = b) \,\&\, \exists c(c \cdot c = b')).$$

EXERCISE 5.7.   If x is a and A(x) is $\exists c(a \cdot c = b)$, what are:

(i) A(0);   (ii) A(b);   and   (iii) A(c)?

t *is free for* x *in* A(x)

DEFINITION 5.8.   We say that a term t *is free for* x *in* A(x) if and only if the *substitution* of t for x in A(x) will not introduce t into A(x) at any place where a (free) variable y of t becomes a bound occurrence of y in A(t). ∎

So, in the formula

$$\exists c(a \cdot c = b) \supset (c \cdot d = b),$$

the terms a, a + c, c + d are all free for c. But in

$$\exists c(a \cdot c = b) \supset \exists d(c \cdot d = b)$$

a, a + c are free for c but c + d is *not*.

EXERCISE 5.8.

(i)   Are the terms 0, b, (b + b), (b + c') free for a in

$$\exists c(c' + a = b)?$$

(ii)   Are $0, q, a + r$ free for b and for r in

$$\neg b = 0 \supset \exists q \exists r (a = b \cdot q + r \ \& \ \exists a(a' + r = b))?$$

(iii)   Are $0', a \cdot c, b$ free for a, b and c in

$$\exists c(c' + 0' = a) \supset \neg (\exists c(a \cdot c = b) \ \& \ \exists c(a \cdot c = b'))? \ \blacksquare$$

The axioms and rules for N

At last we can state the axioms and rules of inference of our formal theory **N**. In 1–8 below A, B, C are formulas of **N**. In 9–13, x is a variable, A(x) is a formula, C is a formula which does not contain x free, and t is a term which is free for x in A(x).

A formula is an *axiom* if it has one of the forms 1(a), 1(b), 3–8, 10, 11, 13 (these are called *axiom schemas*) or if it is one of the formulas 14–21. Each axiom schema is not an axiom itself but it is a 'scheme' which yields an axiom when the letters A, B, C — which are *not* symbols of **N** — are replaced by formulas of **N**. Hence each axiom schema yields a denumerable number of axioms.

The *rules of inference* are given by 2 (usually called *modus ponens*), 9, and 12. A formula is called an *immediate consequence* of one or two formulas if it has the form shown below the line while the other(s) have the form(s) shown above the line in 2, 9, or 12.

Axiom and rules for the formal theory **N**

1(a). $A \supset (B \supset A)$ 2. $\dfrac{A, A \supset B}{B}$

1(b). $(A \supset B) \supset ((A \supset (B \supset C)) \supset (A \supset C))$

3. $A \supset (B \supset A \& B)$ 4(a). $A \& B \supset A$

                            4(b). $A \& B \supset B$

5(a). $A \supset A \lor B$ 6. $(A \supset C) \supset ((B \supset C) \supset (A \lor B \supset C))$

5(b). $B \supset A \lor B$

7. $(A \supset B) \supset ((A \supset \neg B) \supset \neg A)$ 8. $\neg \neg A \supset A$

9. $\dfrac{C \supset A(x)}{C \supset \forall x A(x)}$ 10. $\forall x A(x) \supset A(t)$

11. $A(t) \supset \exists x A(x)$ 12. $\dfrac{A(x) \supset C}{\exists x A(x) \supset C}$

13. $A(0) \, \& \, \forall x(A(x) \supset A(x')) \supset A(x)$

14. $a' = b' \supset a = b$ 15. $\neg a' = 0$

16. $a = b \supset (a = c \supset b = c)$ 17. $a = b \supset a' = b'$

18. $a + 0 = a$ 19. $a + b' = (a + b)'$

20. $a \cdot 0 = 0$ 21. $a \cdot b' = a \cdot b + a$

It is not difficult for the reader to check that under the intended interpretation of the symbols 2 is *modus ponens*, 1–8 (excluding 2) are tautologies, 9–12 are valid axioms and rules of predicate logic, 13 expresses mathematical induction, and 14–21 are truths about identity and numbers. In Chapter 3 when we analysed the informal proofs of elementary number theory in order to extract their underlying assumptions, we found the informal analogues of several of the above axioms and rules, including 1, 2, 4(a), 4(b), 10, 11, 12, 13, 16, 20. The particular choice of axioms and rules which gives us N has been arrived at by extended analysis of the kind we saw in Chapter 2. As will be shown shortly, all the results listed at the end of Chapter 2 are provable in N, or rather their formal analogues are.

Remember, from Definition 4.2, that a *formal proof* of N is a finite sequence of formulas such that each member of the sequence is either an axiom or an immediate consequence of preceding formulas of the sequence and a *theorem* of N is the last member of such a sequence. This makes finding of formal proof in N of (say) $a > 1 \supset \neg(a|b \, \& \, a|b')$ quite different from finding an informal proof of the kind given in Chapter 1. You have to find a *finite* sequence of formulas of N such that each member of the sequence is either an axiom or an immediate consequence of preceding formulas by one of the rules of N.

EXERCISE 5.9. The following sequence is a proof of the formula $a = a$ (taken from Kleene (1952)). Give the justification for each step, e.g. line 1, Axiom 16; line 2, Axiom schema 1(a). (Brackets $\{,\}$, $[,]$ make reading easier.)

1. $a = b \supset (a = c \supset b = c)$

2. $0 = 0 \supset (0 = 0 \supset 0 = 0)$

3. $\{a = b \supset (a = c \supset b = c)\} \supset \{[0 = 0 \supset (0 = 0 \supset 0 = 0)] \supset$
   $[a = b \supset (a = c \supset b = c)]\}$

4. $[0 = 0 \supset (0 = 0 \supset 0 = 0)] \supset [a = b \supset (a = c \supset b = c)]$

5. $[0 = 0 \supset (0 = 0 \supset 0 = 0)] \supset \forall c \, [a = b \supset (a = c \supset b = c)]$

6. $[0 = 0 \supset (0 = 0 \supset 0 = 0)] \supset \forall b \forall c \, [a = b \supset (a = c \supset b = c)]$

7. $[0 = 0 \supset (0 = 0 \supset 0 = 0)] \supset \forall a \forall b \forall c [a = b \supset (a = c \supset b = c)]$

8. $\forall a \forall b \forall c [a = b \supset (a = c \supset b = c)]$

9. $\forall a \forall b \forall c [a = b \supset (a = c \supset b = c)] \supset \forall b \forall c [a + 0 = b \supset (a + 0 = c \supset b = c)]$

10. $\forall b \forall c [a + 0 = b \supset (a + 0 = c \supset b = c)]$

11. $\forall b \forall c [a + 0 = b \supset (a + 0 = c \supset b = c)] \supset \forall c [a + 0 = a \supset (a + 0 = c \supset a = c)]$

12. $\forall c [a + 0 = a \supset (a + 0 = c \supset a = c)]$

13. $\forall c [a + 0 = a \supset (a + 0 = c \supset a = c)] \supset [a + 0 = a \supset (a + 0 = a \supset a = a)]$

14. $a + 0 = a \supset (a + 0 = a \supset a = a)$

15. $a + 0 = a$

16. $a + 0 = a \supset a = a$

17. $a + a$ ∎

EXERCISE 5.10.   Given any formula A, the following sequence is a proof of the formula $A \supset A$. Give the justification for each step.

1. $A \supset (A \supset A)$

2. $\{A \supset (A \supset A)\} \supset \{[A \supset ((A \supset A) \supset A)] \supset [A \supset A]\}$

3. $[A \supset ((A \supset A) \supset A)] \supset [A \supset A]$

4. $A \supset ((A \supset A) \supset A)$

5. $A \supset A$ ∎

Proof-finding made easier

By analysing and systematizing elementary number theory we eventually arrived at the formal theory **N**. At a given stage in the process various truths were found to be derivable from fewer, 'more basic' truths and the latter were preferred in the search for an axiomatization of the theory. We could go further than **N** and find an axiomatization with still fewer and even more basic starting points. For example, axiom schemas 1(a)–8 inclusive could be replaced by the three axiom schemas

$$A \supset (B \supset A)$$

$$(A \supset (B \supset C)) \supset ((A \supset B) \supset (A \supset C))$$

$$(\neg B \supset \neg A) \supset ((\neg B \supset A) \supset B)$$

plus *modus ponens* and the usual definitions for $\lor$, &, and $\sim$. Alternatively we could have stopped the process of analysis sooner.

For certain kinds of investigation of a theory the simpler its axiomatic formulation is the better, but the fewer the axioms and rules the longer are the *formal proofs* within the formal theory and the more difficult they are to find. Our formal theory N has been chosen because it is easily available and well-worked but it carries the process of analysis far enough to make formal proofs very long and quite hard to find. Exercises 5.9 and 5.10 illustrated this very clearly. We now show how to overcome this problem by means of some simple devices. These or similar methods would be available with any other formal theory which had the same purpose. Basically, we show that from the axioms and rules of N we can derive by finitary methods rules of inference and other results which parallel the usual moves of informal number theory, so that the method for showing a given mathematical formula provable in N becomes very similar to the method of proving it informally.

Deduction from assumptions

One practice which is familiar in informal mathematics but for which we do not yet appear to have a formal parallel is deduction from assumptions. In the standard *reductio ad absurdum* proof that $\sqrt{2}$ is irrational one first *assumes* that it is rational. To prove that $a|b \& b|c \supset a|c$ one first *assumes* $a|b \& b|c$. In inductive proofs one *assumes* $A(n)$ in order to derive $A(n + 1)$. There are numerous examples in Chapter 1. We now define a formal analogue of this.

DEFINITION 5.9. *Formal deduction under assumptions (holding variables constant).* A formula B is said to be *formally deducible* from the *assumption formulas* $A_1,\ldots, A_m$ if and only if B is the last member of a finite sequence of formulas such that each one is either one of the assumption formulas $A_1,\ldots, A_m$ or an axiom, or an immediate consequence of preceding formulas of the sequence, *holding variables constant.* This is written $A_1,\ldots, A_m \vdash B$ and is read '$A_1,\ldots, A_m$ yields B'. ∎

This is a generalization of the earlier definition of formal proof, but it contains the phrase 'holding variables constant' which we must now explain. Suppose this phrase were absent, so that a formal deduction was simply a finite list of formulas each being either an assumption formula, or an axiom, or an immediate consequence of preceding formulas, and consider the following 'deduction schema'. Let $A(x)$ be any formula with free variable x and C any axiom not containing x free.

I. 1. A(x)   Assumption formula.

    2. A(x ⊃ (C ⊃ A(x))   Axiom schema 1(a).

    3. C ⊃ A(x)   Immediate consequence of 1 and 2 by *modus ponens*.

    4. C ⊃ ∀xA(x)   Immediate consequence of 3 by rule 9.

    5. C   Axiom.

    6. ∀xA(x)   Immediate consequence of 4 and 5 by *modus ponens*.

In informal mathematics, having derived a result from our initial assumption we are then entitled to conclude that the assumption implies the result, so our formal equivalent would yield

$$\vdash A(x) \supset \forall xA(x)$$

But this will not do. Suppose A(x) is the formula $a + 0 = 1$: we might wish to use this as an assumption formula because we are interested in deriving a result for the case when it is true, when $a = 1$, but clearly it does not follow that $\forall a(a + 0 = 1)$.

Consider another example with A(x) and C as before.

II. 1. A(x)   Assumption formula.

    2. A(x) ⊃ (C ⊃ A(x))   Axiom schema 1(a).

    3. C ⊃ A(x)   Immediate consequence of 1 and 2 by *modus ponens*.

        ⋮   ⋮

    n. ¬A(x) ⊃ ¬C   From 3 by contraposition, $\vdash(A \supset B) \supset (\neg B \supset \neg A)$, a provable tautology.

n + 1. ∃x¬A(x) ⊃ ¬C   Immediate consequence of n by Rule 12.

n + 2.   C   Axiom.

        ⋮   ⋮

m.   ¬∃x¬A(x)   By n + 1, n + 2, contraposition, etc.

        ⋮   ⋮

k.   ∀xA(x)   By m and easily proved equivalence.

The gaps in this finite sequence can easily be filled in uncontroversially but again we cannot conclude that

$$\vdash A(x) \supset \forall xA(x)$$

Since we want the notion of formal deducibility to mirror our informal practice we must block both of these moves.

First we need to define when a formula in a given deduction *depends on* an assumption formula.

DEFINITION 5.10.   In a given deduction with given justifications B *depends on* an assumption formula A if and only if either (i) B is A or (ii) B is an immediate consequence of a formula or formulas which depend on A, i.e. B *depends on* A in a given deduction if and only if by 'tracing back' from B through the formulas of which it was an immediate consequence and through the formulas of which *they* were immediate consequences and so on, we eventually reach the assumption formula A. ∎

We now define *holding variables constant*.

DEFINITION 5.11.   A given deduction with given justifications *holds variables constant* provided rules 9 and 12 are not applied, with x as their variable, to any formula which depends on an assumption formula containing x free. ∎

The reader should check that in deduction schema I above Rule 9 was used to justify line 4 and in deduction schema II Rule 12 was used to justify line n + 1 *without* holding variables constant. If we require deductions from assumptions to hold variables constant these moves are blocked.

EXERCISE 5.11.   In each of the following examples, state which formulas *depend on* the assumption formula and state whether the sequence is a formal deduction from assumptions *holding variables constant*. Give your reasons.

Let A(x) be any formula with free variable x; let y be a different variable.

(i) 1. $A(x)$ — Assumption.

2. $A(x) \supset (\exists x A(x) \supset A(x))$ — Axiom 1(a).

3. $\exists x A(x) \supset A(x)$ — 1, 2, *modus ponens*.

4. $\exists x A(x) \supset \forall x A(x)$ — 3, Rule 9.

(ii) 1. $A(x)$ — Assumption.

2. $A(x) \supset (A(y) \supset A(x))$ — Axiom 1(a).

3. $A(y) \supset A(x)$ — 1, 2, *modus ponens*.

4. $\exists y A(y) \supset A(x)$ — 3, Rule 12.

5. $\exists y A(y) \supset \forall x A(x)$ — 4, Rule 9.

(iii)   1. $\neg A(x)$ Assumption.

2. $\neg A(x) \supset (\forall x A(x) \supset \neg A(x))$ Axiom 1(a).

3. $\forall x A(x) \supset \neg A(x)$ 1, 2, *modus ponens.*

4. $\forall x A(x) \supset A(x)$ Axiom 10.

5. $(\forall x A(x) \supset A(x)) \supset ((\forall x A(x) \supset \neg A(x)) \supset \neg \forall x A(x))$.   Axiom 7.

6. $(\forall x A(x) \supset \neg A(x)) \supset \neg \forall x A(x)$ 4, 5, *modus ponens.*

7. $\neg \forall x A(x)$ 3, 6, *modus ponens.*

8. $\neg \forall x A(x) \supset (\neg A(x) \supset \neg \forall x A(x))$ Axiom 1(a).

9. $\neg A(x) \supset \neg \forall x A(x)$ 7, 8, *modus ponens.*

10. $\exists x \neg A(x) \supset \neg \forall x A(x)$ 9, Rule 12. ∎

These restrictions are all we need; *provided we are careful to hold variables constant*, if from assumption formula A we can formally deduce B then we can formally prove A $\supset$ B. Or more generally, if from a set of assumption formulas Γ and an assumption formula A we can formally deduce B then from Γ we can formally deduce A $\supset$ B. This is the important *Deduction Theorem* which we shortly prove. It is a theorem *about* formal deducibility and formal proof, i.e. it is a metatheorem — it is about proofs in **N** and is not a theorem in **N**. Notice that it is proved by finitary methods because it proceeds by looking at the finite sequence of formulas which establishes Γ, A $\vdash$ B and says how to construct quite mechanically another sequence of formulas which proves that $\Gamma \vdash A \supset B$. The second sequence will be at least three times as long as the first, so this shows how much easier the introduction of formal deduction from assumptions makes the finding of formal proofs.

Before proving the Deduction Theorem, we note some obvious general properties of '$\vdash$' which do not depend on the particular formal theory **N**.

LEMMA 5.1.   For formulas C, D, E, and finite sequences Γ, Δ, of zero or more formulas

(i) $E \vdash E$

(ii) if $\Gamma \vdash E$, then $C, \Gamma \vdash E$

(iii) if $C, C, \Gamma \vdash E$ then $C, \Gamma \vdash E$

(iv) if $\Delta, D, C, \Gamma \vdash E$ then $\Delta, C, D, \Gamma \vdash E$. ∎

Clearly we can add 'redundant' assumption formulas (ii); we can omit duplicate ones (iii); and we can permute them (iv).

The Deduction Theorem

In the following, Γ denotes a (possibly empty) set of formulas, A and B are arbitrary formulas.

THEOREM 5.2. (*The Deduction Theorem*)   If Γ, A $\vdash$ B then $\Gamma \vdash A \supset B$ (holding variables constant).

Proof. Let $B_1, \ldots, B_n$ be a formal deduction of $B(= B_n)$ from Γ, A. We first show that if Γ, A $\vdash B_1$ then $\Gamma \vdash A \supset B_1$ (induction basis). Next we show that if the theorem holds for B_i, $i < k$ then it holds for B_k i.e. if for all i, $i < k$ Γ, A $\vdash B_i$ implies Γ, $\vdash A \supset B_i$ then Γ, A $\vdash B_k$ implies $\Gamma \vdash A \supset B_k$ (induction step).

Basis. If Γ, A $\vdash B_1$ then B_1 is either (i) an axiom, or (ii) in Γ, or (iii) A.

Case (i)   If B_1 is an axiom then by Axiom schema 1(a) $(B_1 \supset (A \supset B_1))$ and *modus ponens* $\vdash A \supset B_1$. Hence, by Lemma 5.1 (ii) $\Gamma \vdash A \supset B_1$.

Case (ii)   If B_1 is in Γ the following sequence establishes $B_1 \vdash A \supset B_1$.

1. B_1 Assumption
2. $B_1 \supset (A \supset B_1)$ Axiom schema 1(a)
3. $A \supset B_1$ 1, 2, and *modus ponens.*

Hence, by Lemma 5.1 (ii) we have $\Gamma \vdash A \supset B_1$.

Case (iii) We saw (p. 50) that $\vdash A \supset A$ so if B_1 is A then $\vdash A \supset B_1$ and hence $\Gamma \vdash A \supset B_1$ again by Lemma 5.1 (ii).

Induction step. We assume the theorem holds for B_i for all i, $i < k$, and show that it holds for B_k. B_k is either (i) an axiom, or (ii) in Γ, or (iii) A, or (iv) an immediate consequence of preceding formulas by rules 2, 9, or 12.

Cases (i)–(iii)   Identical to *Cases* (i)–(iii) in the *Basis* with B_k in place of B_1.

Case (iv)   *Rule 2* (*modus ponens*). B_k is an immediate consequence by Rule 2 of two preceding formulas, B_m and B_j ($= B_m \supset B_k$) where j, $m < k$. By the induction assumption,

$$\Gamma \vdash A \supset B_m \text{ and } \Gamma \vdash A \supset (B_m \supset B_k)$$

and by Axiom schema 1(b),

$$\vdash (A \supset B_m) \supset ((A \supset (B_m \supset B_k)) \supset (A \supset B_k))$$

so by *modus ponens* twice (and easy properties of '$\vdash$') $\Gamma \vdash A \supset B_k$.

Rule 9. B_k ($= C \supset \forall x D(x)$ is an immediate consequence of a preceding formula

$B_m (C \supset D(x))$, $m < k$ by Rule 9 (variables being held constant). There are two possible cases, either (I) B_m depends on A or (II) B_m) does not depend on A.

Case (I).   If B_m depends on A, then (because variables are held constant) A cannot contain x free. Since C does not contain x free nor does (A & C). By the induction hypothesis $\Gamma \vdash A \supset (C \supset D(x))$ so by $\vdash (A \supset (C \supset B)) \supset ((A \& C) \supset B)$ $\Gamma \vdash (A \& C) \supset D(x)$. Thus by rule 9, $\Gamma \vdash (A \& C) \supset \forall x D(x)$. So by $\vdash ((A \& C) \supset B) \supset (A \supset (C \supset B))$, we have $\Gamma \vdash A \supset (C \supset \forall x D(x))$.

Case (II).   If B_m does not depend on A nor does $B_k (= C \supset \forall x D(x))$. So there is a sequence establishing $\Gamma \vdash C \supset \forall x D(x)$. Add to this $\vdash (C \supset \forall x D(x)) \supset (A \supset (C \supset \forall x D(x)))$ Axiom schema 1(a), apply Rule 2, and we have $\Gamma \vdash A \supset (C \supset \forall x D(x))$.

Rule 12.   Similar to Rule 9. In Case (I) use $A \supset (B \supset C) \vdash B \supset (A \supset C)$ twice. ∎

EXERCISE 5.12.   Prove the Deduction Theorem for Rule 12. ∎

Another useful result which follows easily from the deduction theorem is:

THEOREM 5.3.   For $m, p \geqslant 0$

$$\text{If} \quad A_1, \ldots, A_m \vdash B_1$$

$$\vdots$$

$$A_1, \ldots, A_m \vdash B_p \quad and \quad B_1, \ldots, B_p \vdash C$$

$$\text{then} \quad A_1, \ldots, A_m \vdash C. ∎$$

EXERCISE 5.13.   Prove Theorem 5.3.
(*Hint*: use $\vdash (A \supset (B \supset C)) \supset ((A \& B) \supset C).$) ∎

Introduction and elimination rules

There are several other derived rules which, like the Deduction Theorem, make it easier to find theorems of our formal theory **N**. We present some of these now with the names they are usually given by logicians.

In the following list of rules Γ is any set of formulas; A, B, C are arbitrary formulas; x is a variable; A(x) is a formula; and t is a term which is free for x in A(x). We abbreviate 'introduction' to 'intro.' and 'elimination' to 'elim.' throughout.

THEOREM 5.4.   The following rules all hold:

$\supset$-intro.   If Γ, A $\vdash$ B then $\Gamma \vdash A \supset B$. This is easily recognized as the Deduction Theorem.

⊃-elim.   A, A ⊃ B ⊢ B.

&-intro.   A, B ⊢ A & B.

&-elim.   A & B ⊢ A   also   A & B ⊢ B.

∨-intro.   A ⊢ A ∨ B   also   B ⊢ A ∨ B.

∨-elim.   If Γ, A ⊢ C   and   Γ, B ⊢ C   then   Γ, A ∨ B ⊢ C.

¬-intro.   If Γ, A ⊢ B   and   Γ, A ⊢ ¬ B   then   Γ ⊢ ¬A.

¬-elim.   ¬¬A ⊢ A.

∀-intro.   If Γ ⊢ A(x)   then   Γ ⊢ ∀xA(x)   provided Γ does not contain
            a formula with x free.

∀-elim.   ∀xA(x) ⊢ A(t).

∃-intro.   A(t) ⊢ ∃xA(x).

∃-elim.   If Γ, A(x) ⊢ C   then   Γ, ∃xA(x) ⊢ C   provided Γ does not
            contain a formula with x free and C does not contain x free. ∎

EXERCISE 5.14.   Prove them all except ⊃-intro. ∎

The use of ∀-intro and ∀-elim. gives us the following useful rule:

THEOREM 5.5 (*Substitution for an individual variable*).   Let x be a variable,
A(x) a formula, t a term which is free for x in A(x), and Γ not contain x free.
In that case if Γ ⊢ A(x) then Γ ⊢ A(t). ∎

We shall abbreviate this name to 'substitution' since the context will clearly
distinguish it from the use described earlier on p. 47.

Although we do not prove it here, any tautologies or first-order logical truths
which are expressible in the language of N are formal theorems of N. Because
of this, in subsequent work on formal proofs, as we give less and less detail, we
shall occasionally justify steps simply with the words 'Tautology' or 'Logical
Truth'. This will be done only in obvious cases.

We can now proceed to use the results of this chapter to find some theorems
of N.

6
SOME THEOREMS OF N

We have shown that from the *logical* axioms and rules of **N** we can derive formal
rules of inference which parallel the *logical* methods of inference familiar in
informal number theory. This makes proving theorems of **N** similar to finding
informal proofs in elementary number theory. We now display the similarity
in other ways. Beginning with some results involving the formal symbol '=' we
first show that the mathematician's use of = in informal proofs can be copied
in **N**. We then proceed to show that numerous other familiar number-theoretic
results are theorems of **N**. (In what follows we usually abbreviate a·b to ab, and
to facilitate cross-reference, the theorems numbered *85-*180 have the same
numbers as in S. C. Kleene's *Introduction to Metamathematics* (1952).)

Properties of '='

Before we present any of the results of this chapter we give some general advice
about proving theorems of **N**. To show that A is a theorem of **N** first try to
construct an informal proof of A, i.e. prove it as you would if you were doing
informal number theory as in Chapter 1, then try to 'mirror' that proof to show
that $\vdash_N$ A. **N** has been developed so that this should be *relatively* easy. Since we
follow this advice in constructing our own proofs in this chapter, we shall clearly
produce no more full formal proofs like that given in Exercise 5.9 for *100 but
rather we shall take full advantage of the methods developed in the latter part of
Chapter 5: using the methods previously established our proofs will show that
a formal proof of A exists without actually producing it. As in any other branch
of mathematics, results which have been previously proved may be used to ease
the proof of subsequent results in this text. We begin with some properties of
'='.

THEOREM 6.1 (*Properties of* '=')

*100.	$\vdash a = a.$	Reflexivity.
*101.	$\vdash a = b \supset b = a.$	Symmetry.
*102.	$\vdash a = b \,\&\, b = c \supset a = c.$	Transitivity.

*103. $\vdash a = b \supset a' = b'$.   (Axiom 17)

*104. $\vdash a = b \supset a + c = b + c$.

*105. $\vdash a = b \supset c + a = c + b$.

*106. $\vdash a = b \supset ac = bc$.

*107. $\vdash a = b \supset ca = cb$.

*108. $\vdash a = b \supset (a = c \supset b = c)$.   (Axiom 16)

*109. $\vdash a = b \supset (c = a \supset c = b)$.

Proof. (We shall not prove every result in this chapter.)

*100   See Exercise 5.9.

*101

1.	$\vdash a = b \supset (a = a \supset b = a)$	Axiom 16 substituting a for c.
2.	$a = b \vdash a = a \supset b = a$	1, by '$\vdash A \supset B$ implies $A \vdash B$'.
3.	$a = b \vdash a = a$	*100, Lemma 5.1 (ii).
4.	$a = b \vdash b = a$	2, 3, Theorem 5.3.
5.	$\vdash a = b \supset b = a$	4, $\supset$-intro.

*102

1.	$a = b \& b = c \vdash a = b$	&-elim.
2.	$a = b \& b = c \vdash a = b \supset b = a$	*101, Lemma 5.1 (ii).
3.	$a = b, a = b \supset b = a \vdash b = a$	$\supset$-elim.
4.	$a = b \& b = c \vdash b = a$	1, 2, 3, Theorem 5.3.
5.	$a = b \& b = c \vdash b = a \supset (b = c \supset a = c)$	Axiom 16 substituting b for a and a for b, Lemma 5.1 (ii).
6.	$b = a, b = a \supset (b = c \supset a = c) \vdash b = c \supset a = c$	$\supset$-elim.
7.	$a = b \& b = c \vdash b = c \supset a = c$	4, 5, 6, Theorem 5.3.
8.	$a = b \& b = c \vdash b = c$	&-elim.
9.	$b = c, b = c \supset a = c \vdash a = c$	$\supset$-elim.
10.	$a = b \& b = c \vdash a = c$	7, 8, 9, Theorem 5.3.
11.	$\vdash a = b \& b = c \supset a = c$	10, $\supset$-intro.

*104   The proof is by formal induction, $A(c)$ being $a = b \supset a + c = b + c$.

1. $a = b \vdash a + 0 = a$ Axiom 18, Lemma 5.1 (ii).

2. $a = b \vdash a = b$ Lemma 5.1 (i).

3. $a = b \vdash a + 0 = a \,\&\, a = b$ 1, 2, &-intro., Theorem 5.3.

4. $\vdash a + 0 = a \,\&\, a = b \supset a + 0 = b$ *102, substuting $a + 0$
                                                           for a, a for b, and b
                                                           for c.

5. $a = b \vdash a + 0 = b$ 3, 4, Theorem 5.3, etc.

6. $a = b \vdash b = b + 0$ Axiom 18, *101, *modus*
                                                           ponens, and Lemma 5.1 (ii).

7. $a = b \vdash a + 0 = b + 0$ 5, 6, by steps like 1–5.

8. $\vdash a = b \supset (a + 0 = b + 0)$ 7, $\supset$-intro.

9. $a + c = b + c \vdash a + c' = (a + c)'$ Axiom 19, Lemma 5.1 (ii).

10. $a + c = b + c \vdash (a + c)' = (b + c)'$ Substitution in *103.

11. $a + c = b + c \vdash a + c' = (b + c)'$ 9, 10, as in 1–5.

12. $a + c = b + c \vdash (b + c)' = b + c'$ Axiom 19, *101, Lemma
                                                           5.1 (ii).

13. $a + c = b + c \vdash a + c' = b + c'$ 11, 12, as in 1–5.

14. $\vdash a + c = b + c \supset a + c' = b + c'$ 13, $\supset$-intro.

15. $\vdash (a = b \supset (a + c = b + c)) \supset (a = b \supset (a + c' = b + c'))$ 14 by
                                                           $\vdash (A \supset B) \supset ((C \supset A) \supset (C \supset B))$.

16. $\vdash a = b \supset a + c = b + c$ 8, 15, $\forall$-intro., &-intro., then
                                                           formal induction, Axiom 13. ∎

EXERCISE 6.1.   Prove *106 $\vdash a = b \supset ac = bc$. (You may use *105.) ∎

The Replacement Theorem

The preceding results can easily be generalized from variables to arbitrary terms
by repeated applications of the substitution rule given in Theorem 5.5. In
particular, if r and s are terms and $A \sim B$, 'A is equivalent to B', abbreviates
$(A \supset B) \,\&\, (B \supset A)$ then:

*110. $r = s \vdash r' = s',$

*111. $r = s \vdash r + t = s + t;$ *112. $r = s \vdash t + r = t + s;$

*113. $r = s \vdash rt = st$;

*114. $r = s \vdash tr = ts$;

*115. $r = s \vdash r = t \sim s = t$;

*116. $r = s \vdash t = r \sim t = s$;

and these results in turn yield the Replacement Theorem.

THEOREM 6.2 (*Replacement Theorem*).   If t_r is a term containing a specified occurrence of a term r, and t_s is the result of replacing this occurrence by a term s, then

$$r = s \vdash t_r = t_s. \blacksquare$$

EXERCISE 6.2.   Prove Theorem 6.2. (*Hint*: define the *depth* of r to mean the number of operators within whose scope r lies. The proof is by induction on the depth of r in t_r.) ∎

The Replacement Theorem generalizes further in various ways. For example, if A_r is a formula containing a specified occurrence of a term r and A_s is the result of replacing this occurrence by a term s and no variable which occurs in r or s is bound in A_r then $r = s \vdash A_r = A_s$. But Theorem 6.2 is the only replacement result we use.

A final formal identity result which reflects informal mathematical practice and which is very useful is easily proved.

THEOREM 6.3.   If $\Gamma \vdash r = s$ and $\Gamma \vdash s = t$ then $\Gamma \vdash r = t$. (r, s, and t are terms and Γ is any set of formulas, possibly empty.) ∎

EXERCISE 6.3.   Prove Theorem 6.3. ∎

Properties of $+$, $\cdot$, $'$, and 0

We move now to the number theoretic results which are provable in **N** to see if the formal symbols $+$, $\cdot$, $'$, and 0 function like their informal counterparts. Since most of the subsequent proofs are by formal induction, Axiom 13, we begin by illustrating this method with an easy example, the proof that $\vdash 0 + a = a$. (Remember, we may use any previously proved result.) The proof is as follows:

1. $\vdash 0 + 0 = 0$ (Induction base) substituting 0 for a in Axiom 18.

2. $\vdash 0 + a' = (0 + a)'$ Substituting 0 for a and a for b in Axiom 19.

3. $0 + a = a \vdash (0 + a)' = a'$ Replacement Theorem with $r = 0 + a$, $u_r = (0 + a)'$, $s = a$, $u_s = a$.

4. $0 + a = a \vdash 0 + a' = a'$ by 2, 3, and Theorem 6.3.

5. $\vdash 0 + a = a \supset 0 + a' = a'$ 4 and $\supset$-intro.

6. $\vdash \forall a(0 + a = a \supset 0 + a' = a')$ 4, $\forall$-intro., and *modus ponens*

7. $\vdash 0 + 0 = 0 \,\&\, \forall a(0 + a = a \supset 0 + a' = a')$ 1, 6, &-intro., and Theorem 5.3.

8. $\vdash 0 + a = a$ Axiom 13 (substituting $0 + a = a$ for A(a)), 7, and *modus ponens*. ∎

We now give some results which show that the associative, commutative, and distributive laws for $+$ and $\cdot$ are provable. *118 and *122 are used for proving *119 and *123 respectively and we shall need the fact that $\vdash 0 + a = a$ for the proof of *119.

THEOREM 6.4 (*Associative, commutative, and distributive laws for $+$ and $\cdot$*).

*117. $\vdash (a + b) + c = a + (b + c)$.   *121. $\vdash (ab)c = a(bc)$.

*118. $\vdash a' + b = (a + b)'$.   *122. $\vdash a'b = ab + b$.

*119. $\vdash a + b = b + a$.   *123. $\vdash ab = ba$.

*120. $\vdash a(b + c) = ab + ac$. ∎

EXERCISE 6.4.   Prove all these using the following hints: *117 by induction on c; *118 by induction on b; *119 by induction on a using $\vdash 0 + a = a$; *120 by induction on c; *121 by induction on c; *122 by induction on b; *123 first prove $\vdash 0 \cdot a = a$ by induction on a, using any result up to and including *122, then proceed by induction on b. ∎

In what follows we shall abbreviate the term $0'^{\cdots\prime}$ having n successor symbols by n, called a *numeral*. (If $n = 0$ then $n = 0$.) We now show that some elementary properties of 0 and 1 ($0'$) are also provable.

THEOREM 6.5 (*some properties of 0 and 1*).

*124. $\vdash a + 0 = a$.   *125. $\vdash a \cdot 0 = 0$.

*126. $\vdash a + 1 = a'$.   *127. $\vdash a \cdot 1 = a$.

*128. $\vdash a + b = 0 \supset a = 0 \,\&\, b = 0$.   *129. $\vdash ab = 0 \supset a = 0 \vee b = 0$.

*130. $\vdash a + b = 1 \supset a = 1 \vee b = 1$.   *131. $\vdash ab = 1 \supset a = 1 \,\&\, b = 1$.

*132. $\vdash a + c = b + c \supset a = b$.   *133. $\vdash c \neq 0 \supset (ac = bc \supset a = b)$. ∎

EXERCISE 6.5. Prove *127. Prove *132 by induction on c. ∎

Order properties in N

We now give some of the order properties which are provable in **N**. Using the abbreviation 'a < b' for $\exists c(c' + a = b)$ and reading 'a > b' as an abbreviation for b < a, 'a ⩽ b' for a < b ∨ a = b, 'a ⩾ b' for b ⩽ a, and 'a < b < c' for a < b & b < c, we have the order properties we would expect.

THEOREM 6.6 (*order properties*)

*134(a). ⊢ a < b < c ⊃ a < c.   (Transitivity 134(a)–134(d))

*134(b). ⊢ a ⩽ b < c ⊃ a < c.

*134(c). ⊢ a < b ⩽ c ⊃ a < c.

*134(d). ⊢ a ⩽ b ⩽ c ⊃ a ⩽ c.

*135(a). ⊢ a < a'.

*135(b). ⊢ 0 < a'.

*136.   ⊢ 0 ⩽ a.

*137 (= *137_0). ⊢ a = 0 ∨ $\exists b(a = b')$.

*137_1.   ⊢ a = 0 ∨ a = 1 ∨ $\exists b(a = b'')$.

*137_2.   ⊢ a = 0 ∨ a = 1 ∨ a = 2 ∨ $\exists b(a = b''')$.

*138(a). ⊢ a ⩽ b ∼ a < b'.

*138(b). ⊢ a > b ∼ a ⩾ b'.

*139.   ⊢ a < b ∨ a = b ∨ a > b.   (Connexity)

*140.   ⊢ ¬a < a.   (Irreflexivity)

*141.   ⊢ a < b ⊃ ¬a > b.   (Asymmetry)

*142(a). ⊢ a + b ⩾ a.

*142(b). ⊢ b ≠ 0 ⊃ a + b > a.

*143(a). ⊢ b ≠ 0 ⊃ ab ⩾ a.

*143(b). ⊢ a ≠ 0 & b > 1 ⊃ ab > a.

*143(c). ⊢ b ≠ 0 ⊃ a'b > a; hence b ≠ 0 ⊃ $\exists c(cb > a)$.

*144(a). ⊢ a < b ∼ a + c < b + c.

*144(b). ⊢ a ⩽ b ∼ a + c ⩽ b + c.

*145(a). $\vdash c \neq 0 \supset (a < b \sim ac < bc)$.

*145(b). $\vdash c \neq 0 \supset (a \leqslant b \sim ac \leqslant bc)$.

*146(a). $\vdash 0 < b \supset \exists q \exists r(a = bq + r \,\&\, r < b)$.

*146(b). $\vdash a = bq_1 + r_1 \,\&\, r_1 < b \,\&\, a = bq_2 + r_2 \,\&\, r_2 < b \supset q_1 = q_2 \,\&\, r_1 = r_2$.

(Existence and uniqueness of quotient and remainder.) ■

Since most of these results are easily proved we leave them to the interested reader. However, 146(a) and 146(b), the existence and uniqueness of quotient and remainder, were used extensively in Chapter 1, e.g. for Euclid's algorithm, so it is important that they are provable in N. We give their proofs now.

Existence and uniqueness of quotient and remainder

*Proof of *146(a)*: the proof is by induction. Let A(a) be

$$0 < b \supset \exists q \exists r(a = bq + r \,\&\, r < b).$$

1. $\qquad\qquad\qquad \vdash 0 = b \cdot 0 + 0$ 　　Axiom 20, Axiom 18, Replacement Theorem, and Theorem 6.3.

2. $\qquad\qquad 0 < b \vdash 0 = b \cdot 0 + 0 \,\&\, 0 < b$ 　1, Theorem 5.3 and &-intro.

3. $\qquad\qquad 0 < b \vdash \exists q \exists r(0 = bq + r \,\&\, r < b)$ 　2 and ∃-intro. twice.

4. $\qquad\qquad\qquad\qquad \vdash A(0)$ 　　⊃-intro.

5. $0 < b, a = b \cdot q + r \,\&\, r < b \vdash a = b \cdot q + r \,\&\, r < b$

6. $0 < b, a = b \cdot q + r \,\&\, r < b \vdash r < b$ 　　　&-elim.

7. $0 < b, a = b \cdot q + r \,\&\, r < b \vdash r' < b \lor r' = b$ 　　*138(b), etc.

8. $0 < b, a = b \cdot q + r \,\&\, r < b \vdash r' < b \supset a' = bq + r' \,\&\, r' < b$ 　5, Theorem 6.3, Axiom 19, Replacement Theorem, &-elim., &-intro., and ⊃-intro.

9. $0 < b, a = b \cdot q + r \,\&\, r < b \vdash r' < b \supset A(a')$ 　　8, ∃-intro. twice, ⊃-elim., and ⊃-intro.

10. $0 < b, a = b \cdot q + r \,\&\, r < b \vdash r' = b \supset a' = bq + b \cdot 0'$ 　5, Theorem 6.3, Axiom 19, $((b \cdot q + r)' = b \cdot q + r')$, *127 $(b \cdot 0' = b)$ and ⊃-intro.

11. $0 < b, a = b \cdot q + r \ \& \ r < b \vdash r' = b \supset a' = b \cdot (q + 0') + 0 \ \& \ 0 < b$

                                                                     *120 etc.

12. $0 < b, a = b \cdot q + r \ \& \ r < b \vdash r' = b \supset A(a')$ 11 and $\exists$-intro. twice,

                                                                     $\supset$-elim., and $\supset$-intro.

13. $0 < b, a = b \cdot q + r \ \& \ r < b \vdash A(a')$ Proof by cases on 7, 9, 12.

14. $\vdash A(a) \supset A(a')$ 13, $\exists$-elim., $\supset$-intro., and

                                                                $\vdash (A \supset (B \supset (A \supset C)))$

                                                                         $\supset ((A \supset B) \supset (A \supset C)).$

15. $\vdash A(a)$ 4, 14, and induction. ∎

*Proof of *146(b).* The proof of uniqueness is by cases: $q_1 < q_2 \lor q_2 < q_1 \lor q_1 = q_2$. Assume $0 < b, (a = bq_1 + r_1 \ \& \ r_1 < b) \ \& \ (a = bq_2 + r_2 \ \& \ r_2 < b)$. Now $q_1 < q_2 \sim q_2 = q_1 + w'$ (for some w) $\sim bq_2 + r_1 = bq_1 + bw' + r_1 \supset bw' + r_2 = r_1$ by *132. But $bw' \geqslant b$ by *143(a). By contradiction $\neg q_1 < q_2$ and similarly $\neg q_2 < q_1$. Therefore $q_1 = q_2$ which implies $r_1 = r_2$ by *132. The details are left to the reader. ∎

The least number principle

With the order relation $<$ now at our disposal we can express and prove the least number principle (or well-orderedness of the natural numbers). This says that if there exists a natural number x such that $A(x)$ then there exists a least such x, call it y. If x, y, z are distinct variables and $A(x)$ is a formula such that y and z are free for x in $A(x)$ but do not occur free in $A(x)$ then the property of y can be expressed in the formal symbolism by $A(y) \ \& \ \forall z(z < y \supset \neg A(z))$ and the theorem which expresses the least number principle is then:

*149. $\vdash \exists x A(x) \supset \exists y [A(y) \ \& \ \forall z(z < y \supset \neg A(z))]$

Since the proof is quite complicated we outline it here. The reader may either check the details as an exercise or skip the proof if he pleases, simply noting that *149 is needed to prove Euclid's Theorem *161 below.

*Proof of *149.* Let $P(x)$ be

$$\exists y(y < x \ \& \ A(y) \ \& \ \forall z(z < y \supset \neg A(z))$$

and $Q(x)$ be

$$\forall y(y < x \supset \neg A(y)).$$

We first prove $\vdash P(x) \lor Q(x)$ by induction on x by the following steps:

1. $\vdash Q(0)$ $\vdash \neg y < 0, \vdash \neg A \supset (A \supset B)$,

                                                      and $\forall$-intro.

2. $\vdash P(0) \lor Q(0)$ 1 and $\lor$-intro.

3. $P(x) \vdash P(x')$ *135(a) and $\exists$-intro.

4. $P(x) \vdash P(x') \lor Q(x')$ 3 and $\lor$-intro.

5. $\vdash A(x) \lor \neg A(x)$ $\forall$-elim. from
                                                    $\vdash \forall x(A(x) \lor \neg A(x))$.

6. $Q(x), A(x) \vdash P(x')$ *135(a) and $\exists$-intro.

7. $Q(x), A(x) \vdash P(x') \lor Q(x')$ 6 and $\lor$-intro.

8. $Q(x), \neg A(x) \vdash Q(x')$ *138(a), $\forall$-elim., and intro.

9. $Q(x), \neg A(x) \vdash P(x') \lor Q(x')$ 8 and $\lor$-intro.

10. $\vdash P(x) \lor Q(x) \supset P(x') \lor Q(x')$ 5, 7, 9, 4, $\lor$-elim. twice,
                                                    $\supset$-intro.

11. $\vdash P(x) \lor Q(x)$ 2, 10, and induction.

We can now complete the argument:

12. $\vdash P(x') \lor Q(x')$ From 11.

13. $P(x') \vdash \exists y(A(y) \& \forall z(z < y \supset \neg A(z)))$ &-elim.

14. $A(x), Q(x') \vdash A(x)$

15. $A(x), Q(x') \vdash x < x' \supset \neg A(x)$ $\forall$-elim.

16. $A(x), Q(x') \vdash \neg A(x)$ 15, *135(a), and *modus
                                                    ponens*

17. $A(x) \vdash \neg Q(x')$ 14, 16, and $\neg$-intro.

18. $P(x') \lor Q(x'), \neg Q(x') \vdash P(x')$ from $P \lor Q, \neg Q \vdash P$.

19. $A(x) \vdash \exists y(A(y) \& \forall z(z < y \supset \neg A(z))$
                                                    12, 17, 18, 13, and $\lor$-elim.

20. $\vdash \exists x A(x) \supset \exists y[A(y) \& \forall z(z < y \supset \neg A(z))]$
                                                    $\exists$ and $\supset$-intro. ∎

Euclid's Theorem

Before proving Euclid's Theorem we give some results on divisibility. The formal
equivalents of Theorem 1.1 can be easily derived from *152–*155 below. We
use $a|b$, read 'a divides b', as a formal abbreviation for $\exists c(a \cdot c = b)$.

THEOREM 6.7 (*Divisibility results*)

*152.   ⊢ a|ab.

*153.   ⊢ a|a.

*154.   ⊢ a|b & b|c ⊃ a|c.

*155.   ⊢ a > 1 ⊃ ¬(a|b & a|b′).

*156.   ⊢ b ≠ 0 ⊃ (a|b ⊃ 0 < a ⩽ b).

*157.   ⊢ ∃d[d > 0 & ∀b(0 < b ⩽ a ⊃ b|d)].

EXERCISE 6.6.   Prove *153, *154, and *155. (*Hint*: *155 is hard. Use *145(a) and *135(a).) ∎

Using the results of Theorem 6.7 and the least number principle, *149, we now outline the formal proof of Euclid's Theorem, that there are infinitely many primes. We shall use Pr(x), read 'x is prime' as an abbreviation for

$$x > 1 \ \& \ \neg \exists c(1 < c < x \ \& \ c|x)$$

Euclid's Theorem is then stated as follows:

THEOREM 6.8 (*Euclid's Theorem*)

*161.   ⊢ ∃x(Pr(x) & x > a) ∎

Proof. The proof outline of this result is quite long so, again, the reader may skip it or fill in the details as an exercise. We break the proof into several stages. First we show that (it is formally provable that) every number a > 1 has a least divisor y > 1

I   ⊢ a > 1 ⊃ ∃y(y > 1 & y|a & ∀z(z < y ⊃ ¬(z > 1 & z|a))).

For convenience we abbreviate this to a > 1 ⊃ ∃yB(a, y). Secondly we show that if b > 1 is a least divisor of a, then b is prime.

II   ⊢ B(a, b) ⊃ Pr(b).

Finally we show that there is a number d which is the product of all numbers less than or equal to a and the least divisor of d′ is a prime greater than a.

III   ⊢ ∃x(Pr(x) & x > a).

I *To prove* ⊢ a > 1 ⊃ ∃y(y > 1 & y|a & ∀z(z < y ⊃ ¬(z > 1 & z|a))).

Call this a > 1 ⊃ ∃yB(a, y).

1. $\vdash a|a$ *153.

2. $a > 1 \vdash a > 1 \,\&\, a|a$ 1 and &-intro.

3. $a > 1 \vdash \exists x(x > 1 \,\&\, x|a)$ 2 and ∃-intro.

4. $\vdash \exists x(x > 1 \,\&\, x|a) \supset \exists y B(a, y)$ Least number principle with $A(y)$: $y > 1 \,\&\, y|a$.

5. $a > 1 \vdash \exists y B(a, y)$ 3, 4, and *modus ponens.*

6. $\vdash a > 1 \supset B(a, y)$ 5 and ⊃-intro.

II *To prove* $\vdash B(a, b) \supset Pr(b)$.

(Remember that $Pr(b)$ is $b > 1 \,\&\, \neg \exists x(1 < x < b \,\&\, x|b)$.)

1. $B(a, b), 1 < c < b \,\&\, c|b \vdash \forall z(z < b \supset \neg(z > 1 \,\&\, z|a))$ &-elim.

2. $B(a, b), 1 < c < b \,\&\, c|b \vdash c < b \supset \neg(c > 1 \,\&\, c|a)$ ∀-elim.

3. $B(a, b), 1 < c < b \,\&\, c|b \vdash c < b$ 2nd assumption, &-elim., and definition.

4. $B(a, b), 1 < c < b \,\&\, c|b \vdash \neg(c > 1 \,\&\, c|a)$ 2, 3, and *modus ponens.*

5. $B(a, b), 1 < c < b \,\&\, c|b \vdash c|b \,\&\, b|a \supset c|a$ *154.

6. $B(a, b), 1 < c < b \,\&\, c|b \vdash c|b \,\&\, b|a$ &-elim. and intro.

7. $B(a, b), 1 < c < b \,\&\, c|b \vdash c|a$ 5, 6, and *modus ponens.*

8. $B(a, b), 1 < c < b \,\&\, c|b \vdash c > 1$ 2nd assumption, &-elim.

9. $B(a, b), 1 < c < b \,\&\, c|b \vdash c > 1 \,\&\, c|a$ 7, 8, and &-intro.

10. $B(a, b) \vdash \neg(1 < c < b \,\&\, c|b)$ 4, 9, and ¬-intro.

11. $B(a, b) \vdash \forall x \neg(1 < x < b \,\&\, x|b)$ 10 and ∀-intro.

12. $B(a, b) \vdash \neg \exists x(1 < x < b \,\&\, x|b)$

13. $B(a, b) \vdash b > 1$ &-elim.

14. $\vdash B(a, b) \supset Pr(b)$ 12, 13, &-intro., and ⊃-intro.

III *To prove* $\vdash \exists x(Pr(x) \,\&\, x > a)$.

Let $\exists y A(a, y)$ be $\exists y(y > 0 \,\&\, \forall z(0 < z \leqslant a \supset z|y))$.

1. $\vdash A(0, 1)$ $\vdash \neg 0 < z \leqslant 0$ and tautology.

2. $A(a, d) \vdash A(a', da')$ by properties of divisibility.

3. $\vdash \exists y A(a, y)$ ∃-intro. on 1 and 2 and induction.

4.	$A(a, d) \vdash d > 0$	&-elim.
5.	$A(a, d) \vdash d' > 1$	*144(a).
6.	$\vdash d' > 1 \supset \exists y B(d', y)$	by I.
7.	$A(a, d) \vdash \exists y B(d', y)$	5, 6, and *modus ponens.*
8.	$B(d', b) \vdash Pr(b)$	$\supset$-elim. and II.
9.	$B(d', b) \vdash b | d' \& b > 1$	&-elim.
10.	$b > 1 \vdash b | d' \supset \neg b | d$	*155 and tautology.
11.	$B(d', b) \vdash \neg b | d$	9, 10, and *modus ponens*, etc.
12.	$A(a, d) \vdash 0 < b \leqslant a \supset b | d$	& and $\forall$-elim.
13.	$A(a, d), B(d', b) \vdash b > a$	11, 12 after $\supset$-elim., and $\neg$-intro. using *139.
14.	$A(a, d), B(d', b) \vdash Pr(b) \& b > a$	8, 13, and &-intro.
15.	$A(a, d), B(d', b) \vdash \exists x (Pr(x) \& x > a)$	14 and $\exists$-intro.
16.	$A(a, d), \exists y B(d', y) \vdash \exists x (Pr(x) \& x > a)$	15 and $\exists$-elim.
17.	$A(a, d) \vdash \exists x (Pr(x) \& x > a)$	7, 16, and Theorem 5.3.
18.	$\exists y A(a, y) \vdash \exists x (Pr(x) \& x > a)$	17 and $\exists$-elim.
19.	$\vdash \exists x (Pr(x) \& x > a)$	3, 18, and Theorem 5.3. ∎

Equivalents of mathematical induction

We conclude these formal theorems by noting some equivalents of formal induction (Axiom 13).

THEOREM 6.9   If x, y are distinct variables and $A(x)$ is a formula such that y is free for x in $A(x)$ but does not occur free in $A(x)$ then,

*162(a). $\vdash A(0) \& \forall x [\forall y (y \leqslant x \supset A(y)) \supset A(x')] \supset A(x)$.

*162(b). $\vdash \forall x [\forall y (y < x \supset A(y)) \supset A(x)] \supset A(x)$.   (Complete, or course-of-values, induction.)

*163. $\vdash \forall x [A(x) \supset \exists y (y < x \& A(y))] \supset \neg A(x)$.   (Method of infinite descent.) ∎

EXERCISE 6.7. Prove *162(b) (Hard). ∎

Questions about completeness

We have seen that many basic number theoretic results are provable in **N**, in particular, that all the principles listed at the end of Chapter 2 (or rather their formal analogues) are. But the reader will have noticed that some of the results of Chapter 1 are missing. For example, Euclid's Algorithm, the Fundamental Theorem of Arithmetic and the Chinese Remainder Theorem are absent.

Let us consider Euclid's Algorithm. We may write the defining equations for the greatest common divisor function $G(a, b)$ thus:

I   For $b \neq 0$, if $a = b$, $G(a, b) = a$.

II   If $a < b$, $G(a, b) = G(a, b - a)$.

III   If $a > b$, $G(a, b) = G(b, a - b)$.

It is not too difficult to find a formula of **N** which 'expresses' $G(a, b)$. Let us introduce the formal abbreviation $g(a, b)$ as follows:

$$g(a, b) = c \sim c|a \ \& \ c|b \ \& \ \neg \exists d (d > c \ \& \ d|a \ \& \ d|b).$$

Can we now prove the formal equivalents of I, II, and III

(i) $b \neq 0 \vdash a = b \supset g(a, b) = a$.

(ii) $\vdash a + c' = b \supset g(a, b) = g(a, c')$.

(iii) $\vdash a = b + c' \supset g(a, b) = g(b, c')$?

EXERCISE 6.8.   Are these provable in **N**? ■

In this example the search will soon answer the question but if we turn now to the Fundamental Theorem of Arithmetic (FT) the position is different. As yet, we do not have a formula of **N** for expressing the power function, a^b, and without this we cannot express the FT. Where the power is a fixed number, repeated multiplication will serve to express exponentiation. For example, Lagrange's Theorem that every (positive) integer is the sum of four squares is expressed by:

$$\forall n \exists a \exists b \exists c \exists d (n = a \cdot a + b \cdot b + c \cdot c + d \cdot d).$$

However, for the FT we need a form of expression where the powers are variables and at this stage of our work the FT is not expressible in **N**. For present purposes we leave open the question whether FT is expressible and provable in **N** (similarly with the Chinese Remainder Theorem). We shall find in Part II that whether or not these are provable in **N**, **N** is already strong enough for the proof of some surprising results. But first we take a diversion via a related formal system, **D**, which is roughly the theory of addition on the natural numbers, i.e. it is **N** without the symbol · and without axioms 20 and 21. This theory is interesting because, as we shall show, it is *complete*, i.e. all truths expressible in D are theorems of D and it is *decidable*, i.e. for any formula of D we can decide in a finite number of steps whether it is a theorem of D.

7

A COMPLETE THEORY FOR ADDITION

We now demonstrate a standard technique for studying formal theories and we do so with a famous example due to Presburger (1930). We first introduce the notions of *disjunctive normal form* and *prenex normal form*.

Disjunctive normal form

DEFINITION 7.1.   A formula of propositional logic is in *disjunctive normal form* if it is in the form of a disjunction, $D_1 \lor D_2 \lor \ldots \lor D_n$, and each disjunct, D_i, is a conjunction of proposition letters or their negations. ∎

THEOREM 7.1.   Any formula of propositional logic is equivalent to one in disjunctive normal form (d.n.f.).

Proof. We give an example. Suppose P(A, B, C) is a formula containing the three proposition letters A, B, C and let us suppose it has the following truth table:

	A	B	C	P(A, B, C)
	T	T	T	F
(2)	F	T	T	T
	T	F	T	F
	F	F	T	F
	T	T	F	F
(6)	F	T	F	T
(7)	T	F	F	T
	F	F	F	F

Since P(A, B, C) is T only at lines (2), (6), and (7) of the truth table, it is T if and only if (A is F and B is T and C is T) *or* (A is F and B is T and C is F) *or* (A is T and B is F and C is F). In logical notation, P(A, B, C) is equivalent to,

(i) $(\neg A \& B \& C) \vee (\neg A \& B \& \neg C) \vee (A \& \neg B \& \neg C)$

and by easy equivalences this reduces to,

(ii) $(\neg A \& B) \vee (A \& \neg B \& \neg C).$

Both (i) and (ii) are in disjunctive normal form. It should be clear from this example that any formula of propositional logic can be reduced to an equivalent one in disjunctive normal form. For the special case where $P(A, B, C)$ is a tautology or a contradiction it is equivalent to $(A \vee \neg A)$ and $(A \& \neg A)$ respectively. ∎

Prenex normal form

DEFINITION 7.2.   A formula of predicate logic is in *prenex normal form* if it has the form

$$Q_1 x_1 Q_2 x_2 \ldots Q_n x_n (A)$$

where each x_i is a variable, each Q_i is either $\exists$ or $\forall$ and (A) *contains no quantifiers.* ∎

THEOREM 7.2.   Any formula of predicate logic is equivalent to one in prenex normal form.

Proof. We give only an outline. To convert a formula of predicate logic into its equivalent prenex form we need the following easily proved equivalences where x is a variable, $A(x)$ and $B(x)$ are formulas and A, B are formulas which do not contain x free.

*85. $\vdash \neg \forall x A(x) \sim \exists x \neg A(x).$

*86. $\vdash \neg \exists x A(x) \sim \forall x \neg A(x).$

*89. $\vdash A \& \forall x B(x) \sim \forall x (A \& B(x)).$

*90. $\vdash A \vee \exists x B(x) \sim \exists x (A \vee B(x)).$

*91. $\vdash A \& \exists x B(x) \sim \exists x (A \& B(x)).$

*92. $\vdash A \vee \forall x B(x) \sim \forall x (A \vee B(x)).$

*95. $\vdash \forall x (A \supset B(x)) \sim A \supset \forall x B(x).$

*96. $\vdash \forall x (A(x) \supset B) \sim \exists x A(x) \supset B.$

*97. $\vdash \exists x (A \supset B(x)) \sim A \supset \exists x B(x).$

*98. $\vdash \exists x (A(x) \supset B) \sim \forall x A(x) \supset B.$

To convert any formula of predicate logic into an equivalent one in prenex normal form we simply apply *85-*98 repeatedly to move all quantifiers to the

front of the formula. If A, B *do* contain x free we simply change the bound
variable throughout the formula for one which is *not* free in A, B. The proof
that the process terminates is by induction. ∎

EXERCISE 7.1.   Prove *85 and *98. (Hint: for *85 use ⊢¬¬A ⊃ A and
⊢(A ⊃ B) ⊃ (¬B ⊃ ¬A). For *98 use ⊢ A ⊃ B ∼ ¬A ∨ B. The rest are even
easier.) ∎

The formal theory D

We now consider the formal theory obtained from **N** by omitting the two
multiplication axioms, Axioms 20 and 21, and the symbol for multiplication,
·. The resulting system, let us call it **D**, is in effect the formal theory of addition
(only) on the natural numbers. We shall prove: (a) that any true statement
belonging to this theory is a theorem of **D**, i.e. that **D** is *complete;* and (b) that
it is *decidable* for an arbitrary formula of **D**, whether it is a theorem or not.

Since there are no famous unsolved problems (like Fermat's Last Theorem or
Goldbach's Conjecture) involving *only* addition it may not be surprising that **D**
is decidable and complete, but the method of proof and the contrast with the
formal theory **N** are very instructive. The method of proof uses what is called
a 'reduction algorithm'. This is exactly the same in form, though rather more
complicated in detail, as the method of obtaining disjunctive and prenex normal
forms. It is a translation routine which, for a given formula produces at each
step an equivalent formula, this equivalence being formally provable in **D**. The
formulas which result at the end of this procedure, the 'reduced formulas',
are then shown to be complete and decidable and that concludes the proof. Our
proof follows the presentation in Hilbert and Bernays (1934, pp. 366 f.).

Before we can present the algorithm we need three abbreviations.

(i) Some terms in **D** will be of the form

$$(\dots((s+s)+s)+\dots)+s \qquad n \text{ times, where s is any term.}$$

To simplify we shall abbreviate this to s·*n* where *n must* be a numeral. This
does not reintroduce · as in **N**, because a·b, where a·b are both variables,
is not allowed.

(ii) For terms s, t we shall also introduce the abbreviations s < t, s ⩽ t,
etc., defined as in **N** (p. 63).

(iii) For terms s, t, and numeral $n > 1$ we shall define s ≡ t (mod *n*),
(s is congruent to t modulus *n*) by the definition schema

$$s \equiv t \pmod{n} \sim \exists x(s = t + x \cdot n \lor t = s + x \cdot n)$$

(where s, t do not contain x). This yields a different formula for each *n*. E.g. if
$n = 3$ the definition reads without abbreviation

$$s \equiv t \pmod{0'''} \sim \exists x(s = t + ((x + x) + x) \lor t = s + ((x + x) + x))$$

(A congruence $p \equiv q \pmod{n}$, where p, q, n are numbers, is called true if $\mathrm{rm}(p, n) = \mathrm{rm}(q, n)$, otherwise false.)

The Reduction Algorithm

We assume that we begin with a *closed* formula of **D** (i.e. one with no free variables) which contains no abbreviations. We put this into its equivalent prenex normal form $Q_1 x_1, \ldots, Q_n x_n (A)$ where each x_i is a variable of A, each Q_i is $\exists$ or $\forall$ and (A) contains no quantifiers. The reduction algorithm proceeds by showing how to eliminate each of the bound variables and their quantifiers in the order $x_n, \ldots, x_1$. Initially A contains only equalities and their negations but it is 'reduced' by repeated application of the algorithm to a formula containing only inequalities and congruences. The algorithm is now displayed in numbered stages for ease of reference. Notice that in order to avoid breaking up the presentation, all the exercises are collated at the *end* of the chapter (p. 82). The student should read the exposition and then do as many of the exercises as are necessary to ensure understanding.

The instructions of the algorithm are as follows:

(1) If $Q_i x_i$ is $\forall x_i$ replace $\forall x_i$ by $\neg \exists x_i \neg$ so that the quantifier-free expression with which we are dealing, call it $A(x_i)$, is immediately preceded by an existential quantifier.

(2) Put $A(x_i)$ into disjunctive normal form (d.n.f.).

(3) Eliminate equalities and negations by substituting

    (i) $s = t$ by $s < t' \ \& \ t < s'$;   EXERCISE 7.2

    (ii) $\neg s = t$ by $s < t \lor t < s$;   EXERCISE 7.3

    (iii) $\neg s < t$ by $t < s'$; and   EXERCISE 7.4

    (iv) $\neg s \equiv t \pmod{n}$ by $s \equiv t + 0' \pmod{n} \lor s \equiv t + 0'' \pmod{n}$

$$\lor \ldots$$

$$\lor s \equiv t + 0'^{(n-1 \text{ times})} \pmod{n}.$$

EXERCISE 7.5

(4) Restore d.n.f. (Hereafter we omit the subscript x_i and write x.)

(5) In each inequality and congruence gather the xs together so that each side is of the form,

$$x \cdot n + r \quad \text{or} \quad x \cdot n$$

where r is a term which does not contain x but may contain other variables. This we may do by converting a term $s'^{(k \text{ times})}$ into $s + 0'^{(k \text{ times})}$ and using the commutativity of addition.   EXERCISE 7.6

(6) We now ensure that each inequality and congruence has x on only one side. If we have an inequality of the form,

$$x \cdot k + r < x \cdot k + s$$

we reduce it to

$$r < s$$

and one of the form,

$$\left. \begin{array}{l} x \cdot k + r < x \cdot h + s \\[2mm] \text{or} \quad x \cdot h + r < x \cdot k + s \end{array} \right\} \text{where } h > k$$

becomes

$$\left. \begin{array}{l} r < x \cdot p + s \\[2mm] \text{or} \quad x \cdot p + r < s \end{array} \right\} \text{where } p = h - k. \qquad \text{EXERCISE 7.7}$$

If we have a congruence of the form,

$$x \cdot k + r \equiv x \cdot k + s (\mathrm{mod}\ n)$$

we reduce it to

$$r \equiv s (\mathrm{mod}\ n)$$

and one of the form,

$$x \cdot k + r \equiv x \cdot h + s (\mathrm{mod}\ n)$$

becomes

$$x \cdot (k + (h \cdot (n - 1))) \equiv s + (r \cdot (n - 1))\ (\mathrm{mod}\ n)$$

EXERCISE 7.8

(N.B. $a \equiv b (\mathrm{mod}\ n) \sim a + (b \cdot (n - 1)) \equiv b + (b \cdot (n - 1))(\mathrm{mod}\ n) \sim a + (b \cdot (n - 1)) \equiv 0 (\mathrm{mod}\ n)$. Therefore 'subtraction' of a term γ, say, from one side of a congruence modulo n can be achieved by adding $\gamma \cdot (n - 1)$ to both sides of the congruence.)

(7) Every congruence in which x occurs is now of the form,

$$x \cdot p \equiv t (\mathrm{mod}\ n)$$

Initially t need not be a numeral but in that case we change the congruences into equivalents such that there is always a numeral on the right hand side by substituting for,

$$x \cdot p \equiv t (\mathrm{mod}\ n)$$

the n-termed disjunction,

$$(t \equiv 0 \pmod n) \ \& \ x \cdot p \equiv 0 \pmod n))$$

$$\vee \ (t \equiv 0' \pmod n) \ \& \ x \cdot p \equiv 0' \pmod n))$$

$$\vee \dots$$

$$\vee \ (t \equiv 0'^{(n-1)} \pmod n) \ \& \ x \cdot p \equiv 0'^{(n-1)} \pmod n)) \quad \text{EXERCISE 7.9}$$

(8) Restore d.n.f.

After doing this each disjunct of our normal form consists of conjuncts which are either of the form,

$$x \cdot p + r < s,$$

$$\text{or} \ \ r < x \cdot p + s,$$

$$\text{or} \ \ x \cdot p \equiv z \pmod n) \quad \text{where } z \text{ is a numeral,}$$

or the conjuncts do not contain x free.

(9) If a disjunct contains several congruences of the form $x \cdot p \equiv z \pmod n$, say,

$$x \cdot p_0 \equiv z_0 \pmod{n_0} \ \& \dots \& \ x \cdot p_k \equiv z_k \pmod{n_k} \ \& \ C$$

where p_i, z_i, and n_i $(i = 0, \dots, k)$ are numerals and C is all of the disjunct except for the congruences containing x, we substitute for such an expression the disjunction,

$$D_0 \vee D_1 \vee \dots \vee D_{N-1}$$

where $N = n_0 \cdot n_1 \cdot n_2 \cdot \dots \cdot n_k$ and D_j is

$$x \equiv j \pmod N) \ \& \ j \cdot p_0 \equiv z_0 \pmod{n_0} \ \& \dots \& \ j \cdot p_k \equiv z_k \pmod{n_k} \ \& \ C.$$

EXERCISE 7.10

(N.B. If x is a solution and $y \equiv x \pmod n$ then y is a solution (see Exercise 7.10). Therefore if there is a simultaneous solution to the congruences there will be one in the range 0 to $n - 1$, and if x is congruent to a simultaneous solution in this range then x will also be a solution.)

(10) Next, if a disjunct has several inequalities containing x we adjust them so that the terms containing x are identical in each inequality. For example, from

$$x \cdot p_1 + r_1 < s_1, \quad x \cdot p_2 + r_2 < s_2$$

we have equivalently,

$$x \cdot p_1 \cdot p_2 + (r_1 \cdot p_2 + r_2 \cdot p_1) < s_1 \cdot p_2 + r_2 \cdot p_1$$

$$\text{and} \ \ x \cdot p_1 \cdot p_2 + (r_1 \cdot p_2 + r_2 \cdot p_1) < s_2 \cdot p_1 + r_1 \cdot p_2$$

EXERCISE 7.11

We deal similarly with inequalities of the form,

$$x \cdot p_1 + r_1 < s_1, \quad s_2 < x \cdot p_2 + r_2$$

$$\text{or} \quad s_1 < x \cdot p_1 + r_1, \quad s_2 < x \cdot p_2 + r_2.$$

If we use these equivalences repeatedly, each time reducing the number of different terms containing x by one, eventually there will be only one for all of the inequalities in the disjunct, $x \cdot p + t$, say.

(11) Next, we replace all of the inequalities in the disjunct of the form,

$$x \cdot p + t < s_1, \ldots, x \cdot p + t < s_n$$

by an n-termed disjunction where the kth term reads,

$$s_k < s_1' \ \& \ s_k < s_2' \ \& \ \ldots \ \& \ s_k < s_n' \ \& \ x \cdot p + t < s_k$$

(as it is the least upper bound for $x \cdot p + t$ that matters).

EXERCISE 7.12

We deal similarly with inequalities of the form,

$$r_1 < x \cdot p + t, \ldots, r_m < x \cdot p + t$$

replacing them by an m-termed disjunction, the kth term being of the form,

$$r_1 < r_k' \ \& \ \ldots \ \& \ r_m < r_k' \ \& \ r_k < x \cdot p + t.$$

(12) Restore d.n.f.

Now each disjunct contains the variable x in at most three conjuncts; these will include not more than one of the form

$$x \equiv q \pmod{n} \quad \text{(where } q \text{ is a numeral),}$$

not more than one of the form

$$r < x \cdot p + t,$$

and not more than one of the form

$$x \cdot p + t < s$$

where, if both types of inequality occur then the terms in which x occur are the same.

(13) After converting A(x) by processes (5)–(12) to a disjunctive normal form of the given simple form we 'distribute' the existential quantifer which precedes A(x), i.e. we remove it from in front of A(x) and quantify each disjunct which contains x by $\exists x$, ($\exists x (Bx \lor Cx) \sim \exists x Bx \lor \exists x Cx$). Furthermore, for each disjunct thus quantified we put the conjuncts free of x to the left of the $\exists x$.

(14) Now we have only to eliminate x from expressions of the form

$$\exists x(x \equiv q(\bmod n) \ \& \ r < x \cdot p + t \ \& \ x \cdot p + t < s)$$

or the same with one or two conjuncts missing. When $r < x \cdot p + t$ is missing, the inequalities are retrieved by putting

$$0 < x \cdot p + t' \ \& \ x \cdot p + t' < s'.$$

When $x \cdot p + t < s$ is missing, substitute $0 < 0'$ for the whole expression as $\exists x(x \equiv q(\bmod n) \ \& \ r < x \cdot p + t$ is always true. **EXERCISE 7.13**

If the congruence is missing and $p = 1$ we have

$$\exists x(r < x + t \ \& \ x + t < s)$$

for which we write

$$r' < s \ \& \ t < s$$

hence we have eliminated x. **EXERCISE 7.14**

If the congruence is missing and $p > 1$ then we write

$$\exists x(x \equiv 0(\bmod p) \ \& \ r < x + t \ \& \ x + t < s) \quad \textbf{EXERCISE 7.15}$$

(15) Whenever we have a disjunct of the general form,

$$\exists x(x \equiv q(\bmod n) \ \& \ r < x \cdot p + t \ \& \ x \cdot p + t < s)$$

we eliminate p by the equivalence,

$$\exists x(x \equiv q(\bmod n) \ \& \ r < x \cdot p + t \ \& \ x \cdot p + t < s)$$
$$\sim \exists x(x \equiv q \cdot p(\bmod n \cdot p) \ \& \ r < x + t \ \& \ x + t < s)$$

EXERCISE 7.16

Hence we only have to consider expressions of the form

$$\exists x(x \equiv z(\bmod n) \ \& \ r < x + t \ \& \ x + t < s)$$

where z is a numeral while r, s, t need not be and $0 \leqslant z \leqslant n - 1$, because if z is bigger than this we can substitute for it its remainder by division (i.e. the congruence class is equally well defined by its smallest member).

(16) We now substitute for such an expression the disjunct

$$(r < t \ \& \ t + z < s) \lor E_1 \lor E_2 \lor \ldots \lor E_n$$

where each E_i is

$$t < r' \ \& \ r + i < s \ \& \ r + i \equiv t + z(\bmod n).$$

EXERCISE 7.17

This says that *either* (i) $r < t$ and the conditions will be satisfied if and only if the smallest value of x satisfying the congruence also satisfies the inequalities, *or* (ii) $t \leqslant r$ and the conditions will be satisfied if and only if there is a solution for x such that $x + t$ is among the first n successors of r and is smaller than s, $(x = z(\bmod n) \sim x + t \equiv t + z(\bmod n))$.

At this point the 'reduced' formula no longer contains any occurrence of x_i. We now repeat the process for x_{i-1} and so on until all the variables are eliminated.

Completeness and decidability of D

Clearly, any closed formula A of D is either true or false under its intended interpretation.

DEFINITION 7.3.   D is *complete* if any closed formula A of D which is true under its intended interpretation is provable in D, i.e. $\vdash_D A$. Equivalently, D is *complete* if, for any closed formula A of D, either $\vdash_D A$ or $\vdash_D \neg A$. ∎

We have already seen that given any closed formula A of D, we can find an equivalent formula in prenex normal form and the Reduction Algorithn then yields an equivalent formula A_r which has no free variables — i.e. is closed — and has the form of a disjunction of conjunctions of (closed) inequalities and congruences. Since these inequalities and congruences are closed they are each true or false under their intended interpretation.

LEMMA 7.3.   Let M be a closed inequality or congruence of D. Then either M is provable or its negation is provable: either $\vdash_D M$ or $\vdash_D \neg M$. ∎

EXERCISE 7.18.   Prove Lemma 7.3. ∎

THEOREM 7.4 (*Completeness of* D).   Let A be any closed formula of D. Then either $\vdash_D A$ or $\vdash_D \neg A$.

Proof. Let A_r result from applying the Reduction Algorithm to A. By Lemma 7.3, if M is a closed inequality or congruence of A_r either $\vdash_D M$ or $\vdash_D \neg M$. Clearly, the conjunction of a pair of such formulas, M, N, is either provable or its negation is provable: M, N $\vdash_D M \& N$; M, $\neg N \vdash_D \neg(M \& N)$; etc. Again, the disjunction of a pair of disjuncts, P, Q, is either provable or its negation is provable: P, Q $\vdash_D P \vee Q$; P, $\neg Q \vdash_D P \vee Q$; $\neg P, \neg Q \vdash_D \neg(P \vee Q)$. Hence, by induction, $\vdash_D A_r$ or $\vdash_D \neg A_r$. Thus, by the equivalence of A and A_r, either $\vdash_D A$ or $\vdash_D \neg A$. ∎

This concludes the proof that D is complete. It is also possible to show that D is *consistent* but the proof is long so we omit it here. We assume D is consistent for our next result.

THEOREM 7.5 (*Decidability of* D).   If A is an arbitrary formula of D it is decidable whether $\vdash_D A$ or not.

Proof. A is provable if and only if its closure A_c is provable (by $\forall$-intro. and $\forall$-elim.). The Reduction Algorithm not only showed that $\vdash_D A_c$ or $\vdash_D \neg A_c$ but also which was the case. ∎

Properties of congruences

We conclude this chapter with some easy properties of congruences which are provable in **D** and which may be used in the Exercises.

THEOREM 7.6 (*Properties of congruences*). The following are theorems of **D**:

(i) $\vdash a \equiv a (\bmod\, n)$;

(ii) $\vdash a \equiv b (\bmod\, n) \sim b \equiv a (\bmod\, n)$;

(iii) $\vdash a \equiv b (\bmod\, n) \,\&\, b \equiv c (\bmod\, n) \supset a \equiv c (\bmod\, n)$;

(iv) $\vdash a \equiv b (\bmod\, n) \sim a + c = b + c (\bmod\, n)$;

(v) $\vdash a \cdot n \equiv 0 (\bmod\, n)$.

For any particular n we have (vi) and (vii),

(vi) $\vdash a \equiv 0 (\bmod\, n) \lor a \equiv 1 (\bmod\, n) \lor \ldots \lor a \equiv n - 1 (\bmod\, n)$;

(vii) $\vdash a \equiv b (\bmod\, n) \lor a \equiv b + 1 (\bmod\, n) \lor \ldots \lor a \equiv b + n - 1 (\bmod\, n)$;

(viii) $\vdash a \equiv 0 (\bmod\, n) \supset a = 0 \lor a \geqslant n$;

(ix) $\vdash a \equiv b (\bmod\, n \cdot m) \supset (a \cdot k \equiv c (\bmod\, n) \sim b \cdot k \equiv c (\bmod\, n))$.

Proof. We give only sketches. A justification like *111$_D$ refers to *111 given earlier, but indicates that it also holds in **D**. Remember that

$$a \equiv b (\bmod\, n) \sim \exists x (a = x \cdot n + b \lor b = x \cdot n + a)$$

where n is a numeral and $n > 1$.

(i) From $\vdash_D a = 0 \cdot n + a$.

(ii) From $\vdash (a = x \cdot n + b \lor b = x \cdot n + a) \sim (b = x \cdot n + a \lor a = x \cdot n + b)$.

(iii) Proof by Cases, e.g.

$$a = x \cdot n + b, \, b = y \cdot n + c \vdash a = (x + y) \cdot n + c.$$

(iv) Proof by Cases both ways, e.g.

$$a = x \cdot n + b \vdash a + c = x \cdot n + b + c \qquad *111_D$$

$$\text{and} \quad a + c = y \cdot n + b + c \vdash a = y \cdot n + b \qquad *132_D.$$

(v) From $\vdash a \cdot n = a \cdot n + 0.$

(vi) Proof is by induction on a. The basis comes from $\vdash 0 \equiv 0 (\bmod n)$, from (i). The induction step is by cases from,

$$a \equiv 0(\bmod n) \vdash a' \equiv 1(\bmod n) \qquad\qquad\qquad \text{(iv)}$$
$$\vdots$$
$$a \equiv n - 2(\bmod n) \vdash a' \equiv n - 1(\bmod n) \qquad\qquad \text{(iv)}$$
$$a \equiv n - 1(\bmod n) \vdash a' \equiv 0(\bmod n) \qquad\qquad \text{(iv), (v), and (iii).}$$

(vii) Proof is by induction on b. The basis is from (vi)

$$\vdash a \equiv 0(\bmod n) \vee a \equiv 0 + 1(\bmod n) \vee \ldots \vee a \equiv 0 + n - 1(\bmod n).$$

The induction step is by cases from,

$$a \equiv b(\bmod n) \vdash a \equiv b' + n - 1(\bmod n) \qquad \text{(v), (iv), (ii), and (iii).}$$
$$a \equiv b + 1(\bmod n) \vdash a \equiv b'(\bmod n) \qquad\qquad \text{Axiom } 18_D, 19_D.$$
$$\vdots$$
$$a \equiv b + n - 1(\bmod n) \vdash a \equiv b' + n - 2(\bmod n).$$

(viii) Proof by Cases from,

$$0 = x \cdot n + a \vdash a = 0 \vee a \geqslant n \qquad\qquad \text{by } *128_D \text{ and tautology.}$$
$$a = x \cdot n + 0 \vdash a = 0 \vee a \geqslant n \qquad\qquad \text{by cases from } *137_D.$$

(ix) Proof by Cases both ways, e.g.

$$a = d \cdot n \cdot m + b, c = e \cdot n + a \cdot k \vdash c = (e + d \cdot k \cdot m) \cdot n + b \cdot k \text{ etc. } \blacksquare$$

Exercises for Chapter 7

All the following exercises relate to the formal system **D**. Except for the first and last they correspond to a step in the Reduction Algorithm just presented. Some of the easier exercises are not provided with solutions. You should do only as many problems as you need in order to understand Presburger's result.

7.1. Prove $\vdash \neg \forall x A(x) \sim \exists x \neg A(x)$ and $\vdash \exists x (A(x) \supset B) \sim \forall x A(x) \supset B.$ $\blacksquare$

7.2. Prove $\vdash a = b \sim a < b' \& b < a'.$
(Hint: prove $\vdash a = b \supset a < b'$ by Axioms $17_D, 18_D, 19_D$ then
$\vdash \neg a = b \supset a < b \vee b < a$ by $*139_D$ and then $\vdash a < b \supset \neg b < a'$ by $\neg$-elim.) $\blacksquare$

7.3. Prove $\vdash \neg a = b \sim a < b \vee b < a.$
(Hint: $\supset$ by $*139_D$; $\subset$ by Cases. Where we have to prove $\vdash A \sim B$, we write '$\supset$' for $A \supset B$ and '$\subset$' for $B \supset A$.) $\blacksquare$

7.4. Prove $\vdash \neg a < b \sim b < a'$.
(Hint: $\supset$ by *139$_D$ and Cases; $\subset$ by $\neg$-elim.) ∎

7.5. Prove $\vdash \neg a \equiv b(\bmod n) \sim a \equiv b + 1(\bmod n) \vee a \equiv b + 2(\bmod n) \vee \ldots \vee$
$a \equiv b + n - 1(\bmod n)$.
(Hint: $\supset$ by Theorem 7.6 (vii) and tautology; $\subset$ by Cases using Theorem 7.6
(ii), (iii), (iv), and (viii) to get a contradiction for each case.) ∎

7.6. Show that if t is a term containing the variable x, there is a term of the
form $x \cdot n + r$ where r does not contain x such that $\vdash t = x \cdot n + r$.
(Hint: for any 'subterm' in t of the form $s'^{(n \text{ times})}$ convert this into $s'^{(n-1)} + 0'$
repeatedly, then use associativity and commutativity of addition. ∎

7.7. Check that (i)–(iii) below can be proved from

$$*144(a)_D \quad \vdash a < b \sim a + c < b + c$$

and $*145(a)_D \quad \vdash c \neq 0 \supset (a < b \sim ac < bc)$;

(i) $\vdash x \cdot k + r < x \cdot k + s \sim r < s$.

(ii) $\vdash h = k + p \supset (x \cdot k + r < x \cdot h + s \sim r < x \cdot p + s)$.

(iii) $\vdash h = k + p \supset (x \cdot h + r < x \cdot k + s \sim x \cdot p + r < s)$. ∎

7.8. Check that the following can be proved from the properties of congruences:

(i) $\vdash x \cdot k + r \equiv x \cdot k + s(\bmod n) \sim r \equiv s(\bmod n)$;

(ii) $\vdash x \cdot k + r \equiv x \cdot h + s(\bmod n) \sim x \cdot (k + (h \cdot (n - 1))) + r \equiv s(\bmod n)$;

(iii) $\vdash x \cdot p + r \equiv s(\bmod n) \sim x \cdot p \equiv s + (r \cdot (n - 1))(\bmod n)$. ∎

7.9. Prove $\vdash x \cdot p \equiv t(\bmod n) \sim [t \equiv 0(\bmod n) \ \& \ x \cdot p \equiv 0(\bmod n) \vee t \equiv 1(\bmod n)$
$\& \ x \cdot p \equiv 1(\bmod n) \vee \ldots \vee t \equiv n - 1(\bmod n) \ \& \ x \cdot p \equiv n - 1(\bmod n)]$.
(Hint: $\supset$ by Cases, Theorem 7.6 (vi), (iii); $\subset$ by Cases and Theorem 7.6 (iii).) ∎

7.10. Prove $\vdash x \cdot p_0 \equiv z_0(\bmod n_0) \ \& \ x \cdot p_1 \equiv z_1(\bmod n_1) \ \& \ \ldots \ \& \ x \cdot p_k \equiv z_k(\bmod n_k)$
$\sim [x \equiv 0(\bmod N) \ \& \ 0 \cdot p_0 \equiv z_0(\bmod n_0) \ \& \ \ldots \ \& \ 0 \cdot p_k \equiv z_k(\bmod n_k)]$
$\vee$
$\vdots$
$\vdots$
$\vee x \equiv N - 1(\bmod N) \ \& \ (N - 1) \cdot p_0 \equiv z_0(\bmod n_0) \ \& \ \ldots \ \& \ (N - 1) \cdot p_k \equiv z_k(\bmod n_k)]$

where $N = n_0 \cdot n_1 \ldots n_k$.
(Hint: $\supset$ prove by Cases from Theorem 7.6 (vi) (applied to the numeral N) using
Theorem 7.6 (ix) repeatedly to prove each case along with &-elim., &-intro.,
and $\vee$-intro.; $\subset$ prove by Cases using Theorem 7.6 (ix) repeatedly to prove each
case.) ∎

7.11. Check $\vdash (x \cdot p_1 + r_1 < s_1 \ \& \ s_2 < x \cdot p_2 + r_2) \sim (x \cdot p_1 \cdot p_2 + r_1 \cdot p_2 + r_2 \cdot p_1 < s_1 \cdot p_2 + r_2 \cdot p_1 \ \& \ s_2 \cdot p_1 + r_1 \cdot p_2 < x \cdot p_1 \cdot p_2 + r_2 \cdot p_1).$ ∎

7.12. Check $\vdash x < s_1 \ \& \ x < s_2 \ \& \ ... \ \& \ x < s_n \sim (s_1 < s_1 + 1 \ \& \ ... \ \& \ s_1 < s_n + 1 \ \& \ x < s_1$

$\vee$

$\vdots$

$\vee \, s_n < s_1 + 1 \ \& \ ... \ \& \ s_n < s_n + 1 \ \& \ x < s_n).$

(Hint: $\supset$ by the least number principle; $\subset$ by cases.) ∎

7.13. Prove $\vdash \exists x (x \equiv q \pmod{n} \ \& \ a < x \cdot p + t).$
(Hint: Let $x = a' \cdot n + q$ and $n > 1$.) ∎

7.14. Prove $\vdash \exists x (a < x + t \ \& \ x + t < b) \sim a' < b \ \& \ t < b.$ ∎

7.15. Prove $\vdash p > 1 \supset (\exists x (a < x \cdot p + t < b) \sim \exists x (x \equiv 0 \pmod{p} \ \& \ a < x + t < b))$
(Hint: $\vdash a < x \cdot p + t < b \sim x \cdot p = x \cdot p + 0 \ \& \ a < x \cdot p + t < b.$) ∎

7.16. Prove $\vdash \exists x (x \equiv q \pmod{n} \ \& \ a < x \cdot p + t < b) \sim \exists x (x \equiv q \cdot p \pmod{n \cdot p} \ \& \ a < x + t < b).$
(Hint: $\vdash x = k \cdot n + q \ \& \ a < x \cdot p + t < b \sim x \cdot p = k \cdot n \cdot p + q \cdot p \ \& \ a < x \cdot p + t < b.$) ∎

7.17. Prove $\vdash z < n \supset \{ \exists x (x \equiv z \pmod{n} \ \& \ a < x + t < b) \sim [(a < t \ \& \ t + z < b) \vee (t < a' \ \& \ a + 1 < b \ \& \ a + 1 \equiv t + z \pmod{n})$

$\vee$

$\vdots$

$\vee \, (t < a' \ \& \ a + n < b \ \& \ a + n \equiv t + z \pmod{n})] \}.$

(Hint: $\vdash a < t \vee t < a'; x \equiv z \pmod{n} \sim x + t \equiv t + z \pmod{n};$
  $\vdash x + t \equiv a + 1 \pmod{n} \vee ... \vee x + t \equiv a + n \pmod{n};$
 and $\vdash c \equiv d \pmod{n} \ \& \ d < n \supset d < c.$) ∎

7.18. Show that any given closed inequality or congruence is either provable or its negation is provable.
(Hint: for inequalities assume they are not provable. For congruences consider only $k \equiv 0 \pmod{n}$ and use the result for inequalities with Theorem 7.6.) ∎

PART II

**Computability,
incompleteness,
and undecidability**

8

INTRODUCING REGISTER MACHINES

We now introduce a new subject, that of *computability*. It will have been clear
that some of the previous exercises required ingenuity and mathematical inventiveness for their solution, e.g. the proof in Chapter 1 that if p is prime and
$p|ab$ then $p|a$ or $p|b$. Others required only the application of 'mechanical'
routines, either already assumed or presented in the text, e.g. the truth-table
method for deciding if a formula is a tautology.

Intuitions about computability

Beginning with the simplest examples, it is obvious that it is a purely 'mechanical'
matter to calculate for arbitrary numbers a, b, the values of $a + b$ and $a \cdot b$, or to
find q and r such that $a = b \cdot q + r$. It is clear too that Euclid's Algorithm is a
purely mechanical procedure for finding the highest common factor of a, b.
The method of Eratosthenes' sieve, for deciding if a is prime, is also a mechanical procedure. All these and many other examples are intuitively clear even without any precise definition of 'mechanical' procedure.

EXERCISE 8.1.   Give mechanical instructions (describe a 'machine') for find-
ing the q, r such that $a = b \cdot q + r$ and $0 \leqslant r < b$, where a, b are arbitrary natural
numbers.

 (i) How much paper will you (your machine) need in order to cope with *any*
     numbers a, b?
(ii) Do your instructions deal with all possible cases (e.g. $b \geqslant a$, $b = 0$)?
(iii) Are your instructions *precise*? Is there any uncertainty about when they
      apply, which *you* would have to judge but which would leave your machine
      whirring on indefinitely, unable to decide what to do next?
(iv) How many basic instructions have you used? ∎

   Non-numerical examples are easy to find too. Examples include the truth-
table method for deciding if a formula is a tautology, the process used in the
proof of the deduction theorem for converting a sequence which shows
$\Gamma, A \vdash B$ into one which shows $\Gamma \vdash A \supset B$, and the method for deciding if a
formal expression of **N** is a *formula*. In fact we shall shortly indicate how to

construct a machine which does the last of these.

It is not easy to say what all these examples of 'mechanical' processes have in common, except that in each case there is some *finite* set of basic *instructions* which apply to any appropriate argument (either numbers or formal expressions in the example above) and which achieve their effect by repeated application a *finite* number of times. But such an observation is too general to be of any mathematical use. Restricting our attention for the moment to numerical functions, we could carry through a program of listing the basic instructions which seem appropriate to each of our numerical examples above, — analysing and systematizing them in a way parallel to the method of Chapter 2. This 'discovery' method could then lead us (though slowly!) to a mathematically usable characterization of the computable numerical functions just as we were led to a *formal* number theory by digging out the logical and mathematical assumptions underlying informal number theory. But this would take too long. Instead, we shall again take advantage of discoveries made by previous workers in the field and present straight away a very natural, elegant, and simple characterization of the computable functions, due to Shepherdson and Sturgis (1963).

In the early 1930s there was strong interest among mathematicians in characterizing the computable functions. Work progressed independently in a number of places and researchers like Church, Gödel, Kleene, Markov, Post, and Turing produced apparently different answers. (Most of the basic papers are in Davis (1965).) Remarkably, all these turned out to be *equivalent* — any function reckoned computable by one of these definitions is also computable by all the others. The definition we present is also equivalent to each of these. We shall return to this point when we discuss Church's Thesis at the end of this chapter.

The unlimited register ideal machine described

We shall characterize the computable functions as those functions which are computable by a particular machine, an *unlimited register ideal machine*, or 'R-machine' for short. This machine is very like an ordinary digital computer: it obeys a *program*, which contains a *finite* number of *instructions*; it has *stores* or *registers* which can contain numbers; and it acts on the contents of its registers in a succession of simple steps. The steps occur at what are called successive *moments* of time. The action it takes at moment $t + 1$ is always *wholly determined* by the program and the register contents at moment t. Of course, only the contents of registers *referred to in the program* can determine what the machine does. In contrast to an ordinary digital computer, the R-machine is an *ideal* machine in two respects. First, a given register can hold an arbitrarily large positive integer. Second, the machine has an *unlimited* supply of registers in which to store its 'information'. Actual computers, although they can be enlarged or extended, always have limitations in these respects.

We now proceed to a detailed description of the R-machine. It has a potentially infinite sequence of *registers* $R_1, R_2, R_3...$, and an *instruction counter* which we shall call R_0.

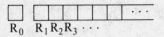

$$R_0 \quad R_1 R_2 R_3 \cdots$$

A register, or the instruction counter, may either contain *any* positive integer or be empty. We let r_i stand for the contents of R_i. If R_i is empty it is considered to contain the number 0. At any moment only a *finite* number of registers may be non-empty.

A *program* is a *finite* sequence of instructions $I_0,..., I_h$. Instruction I_k is called the kth instruction of the program. The machine carries out these instructions consecutively, in the order $I_0, I_1, I_2,...$, unless instructed by a 'jump' instruction to do otherwise. Each *instruction* has one of the forms (1)–(4).

(1) S_m, where $m \geqslant 1$ (read, 'successor the contents of R_m').

To obey this instruction the machine replaces the contents r_m of R_m by its successor, $r_m + 1$, and adds 1 to the contents of R_0, the instruction counter. The contents of all other registers are left unchanged.

(2) Z_m, where $m \geqslant 1$ (read, 'zero the contents of R_m').

To obey this instruction the machine writes 0 in R_m (making R_m *empty*) and adds 1 to the contents of R_0. The contents of all other registers are left unchanged.

(3) $J_{i,j,k}$ where $i, j \geqslant 1$ and $k \geqslant 0$ (read, 'jump').

Let the contents of R_i and R_j at moment t be r_i and r_j. To obey this instruction the machine proceeds thus: if $r_i = r_j$ then put k in R_0 (i.e. 'jump' to instruction I_k); if $r_i \neq r_j$ then add 1 to the contents of R_0 (i.e. go to the next instruction in the program). In either case the contents of all registers except the instruction counter remain unchanged.

(4)   STOP. To obey this instruction the machine stops.

Every program is supposed to contain just one STOP instruction as its last one, I_h.

We assume that the machine *starts* a computation with 0 in the instruction counter; this counter then contains successively (i.e. at successive *moments*) the number of the instruction which is to be obeyed at that moment.

Once started, the machine may or may not stop, or 'halt' as we shall often say. For example, the program,

$$I_0 = J_{1,1,0}$$

$$I_1 = \text{STOP}$$

will send the machine whirring on indefinitely, never reaching instruction $I_1 = \text{STOP}$.

EXERCISE 8.2.   Describe what the following program does.

$$I_0 = J_{1,2,3}$$
$$I_1 = S_1$$
$$I_2 = J_{1,1,0}$$
$$I_3 = Z_1$$
$$I_4 = \text{STOP} \ \blacksquare$$

It can be difficult to devise a program capable of performing a given operation. There are two ways to simplify this task: (i) by drawing a flow diagram first; and (ii) by constructing complex programs out of simpler ones. We show how to draw flow diagrams first.

Drawing flow diagrams

We begin with a trivial example. Suppose we want a program which will copy the contents of R_1 into R_3 leaving all registers other than R_3 unchanged. In particular, R_1 will retain its original number. The following is a suitable flow diagram:

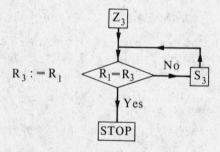

The abbreviation '$R_3 := R_1$' is read thus: 'copy the contents of R_1 into R_3'. The flow diagram is to be read as follows:

(i) Start at the top left corner by obeying the instruction Z_3. A rectangular box means 'Obey the instruction(s) in the box.'

(ii) Follow the arrows to

A diamond-shaped box represents a *question*. In this case, the question is 'Is the number contained in R_1 identical with the number contained in R_3?'

(iii) The two arrows leading from the diamond correspond to the answers 'Yes' and 'No'. If $r_1 \neq r_3$ the arrow is followed to $\boxed{S_3}$, and then again to

$$\langle R_1 = R_3 \rangle$$

We continue around this 'loop', as it is called, until the answer is 'Yes, $r_1 = r_3$'.

(iv) When $r_1 = r_3$, the arrow is followed to STOP.

It is easy to see that the flow diagram, $R_3 := R_1$, displays the required action. To convert it into a program we follow the arrows as before, using one jump instruction to leave a diamond-shaped box and another to complete a loop:

$$I_0 = Z_3 \qquad \text{Zero } R_3.$$

$$I_1 = J_{1,3,4} \qquad \begin{array}{l} \text{If } r_1 = r_3 \text{ go to instruction 4.} \\ \text{If } r_1 \neq r_3 \text{ go to next instruction.} \end{array}$$

$$I_2 = S_3$$

$$I_3 = J_{1,1,1} \qquad \text{If } r_1 = r_1 \text{ go to instruction 1.}$$

$$I_4 = \text{STOP}. \qquad \text{Stop.}$$

$I_3 = J_{1,1,1}$ may seem odd, but it is simply a device within the instructions the machine can obey. It enables us to go to whichever instruction we choose regardless of the contents of the registers. It functions in this way because it is always true that $r_1 = r_1$.

Another, more interesting example, is a program which will compute the function $m + n$. Suppose we put the arguments m and n in registers R_1 and R_2 respectively, and suppose we wish to have the solution, $m + n$, in register R_1. The following flow diagram will suffice for this. We shall refer to it by means of the abbreviation, $R_1 := R_1 + R_2$.

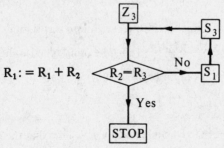

We first make R_3 contain zero. Remembering that R_2 contains n we then ask
whether $n = 0$. If $n = 0$, we stop; our answer, $m + 0 = m$, is already in R_1.
If $n \neq 0$ we add 1 to both r_1 and r_3. We keep adding 1 to both r_1 and r_3, i.e.
we keep going round the loop, until $r_2 = r_3$, by which moment we have added
1 n times to both r_1 and r_3. Then we stop and the answer, $m + n$, is in R_1.
R_3 has been used throughout as a counter, counting the number of times we
have added 1 (or gone round the loop). Proceeding as before the convert this
flow diagram into a program we get the following:

$$I_0 = Z_3$$

$$I_1 = J_{2,3,5}$$

$$I_2 = S_3$$

$$I_3 = S_1$$

$$I_4 = J_{1,1,1}$$

$$I_5 = \text{STOP}$$

Clearly, this program could be transformed into a program, $R_i: = R_i + R_j$, which
adds the contents of any two distinct registers R_i and R_j. Similarly, our previous
program will generalize to $R_j: = R_i$, which copies the contents of R_i into R_j for
any i, j where $i \neq j$.

EXERCISE 8.3.   Give the flow diagram and program $R_j: = R_i$. ■

Building from 'subprograms'

If simple programs are to be available as 'subprograms' for building more complex
ones, we shall obviously need them in a generalized form for arbitrary registers
R_i, R_j. This is how we shall develop further examples. However, we must first
note three 'complications'.

Suppose that in the course of some more complex flow diagram we have:

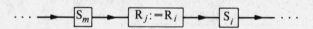

Obviously, when converting the flow diagram into instructions, the STOP
instruction at the end of $\boxed{R_j: = R_i}$ must be omitted. Also the instruction
numbers for $\boxed{R_j: = R_i}$ cannot begin with I_0. It is easy to make the necessary
number changes.

EXERCISE 8.4.   For the example just given assume that the instruction

S_m is I_{12}. Give $I_{13}, \ldots, I_{17}$. ■

We illustrate the third complication by reference to the program for addition.

EXERCISE 8.5.

(i) Does the program $R_j := R_i$ use any register other than R_j and R_i. Assume $i \neq j$.

(ii) Does the program $R_1 := R_1 + R_2$ use any registers other than R_1 and R_2? ■

If we are to add the contents of R_i and R_j in the course of some more complex program, we must be careful to ensure that the 'working' is done in a register which is not already being used for some other purpose. To ensure this, let R_k be the highest-numbered register in use before the addition (sub)program is written out in full. We use R_k to obtain a generalized flow diagram and program: $R_i := R_i + R_j$, where we assume $i \neq j$.

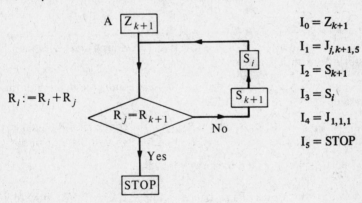

$$I_0 = Z_{k+1}$$
$$I_1 = J_{j, k+1, 5}$$
$$I_2 = S_{k+1}$$
$$I_3 = S_i$$
$$I_4 = J_{1,1,1}$$
$$I_5 = \text{STOP}$$

Multiplying the contents of R_i and R_j

Turning now to our third example, a program for computing $m \cdot n$, we shall use both previous examples as subprograms. This will illustrate how to deal with the problems just mentioned.

Suppose we put the arguments m, n in registers R_i, R_j respectively and suppose we want the solution, $m \cdot n$, in R_i. The following is a suitable flow diagram; again $i \neq j$.

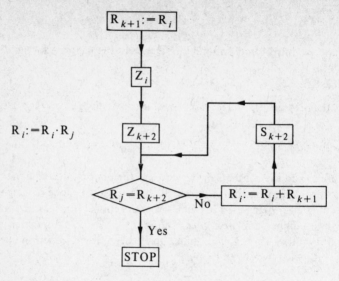

The diagram is read thus:

(i)   Copy the contents of R_i into R_{k+1}. R_{k+1}, previously *unused*, is simply storing the initial contents of R_i.

(ii)   Zero R_i and R_{k+2}. R_{k+2}, previously *unused*, counts the number of times m is added to the contents of R_i.

(iii)   Does $r_{k+2} = n$? If No, add m (in R_{k+1}) to the contents of R_i, and put the answer in R_i. Then successor R_{k+2}. Continue around the loop until $r_{k+2} = n$.

(iv)   When $r_{k+2} = n$, stop and the answer is in R_i.

We now show by means of a particular example, $R_1 := R_1 \cdot R_2$, how to convert such a flow diagram into a program. At the first stage, only registers R_1 and R_2, containing the arguments m and n, are in use, so $k = 2$. It is a routine matter to produce the first-stage flow diagram from the general case. At the second stage, we have to expand

$$\boxed{R_3 := R_1}$$

into a flow diagram which uses only basic instructions. R_4 is the highest number numbered register now in use, so $k = 4$. At the third stage we have to expand

$$\boxed{R_1 := R_1 + R_3}$$

again, $k = 4$. Finally, to write out the program we simply follow the arrows, going round loops before continuing. The reader should check the details.

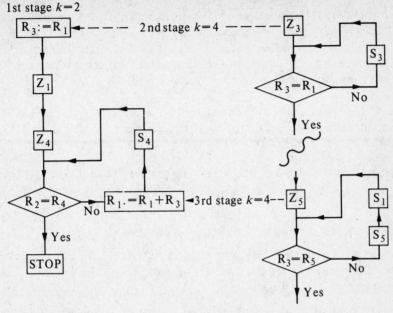

1st stage $k=2$

$R_3 := R_1$ ◄ — — — 2nd stage $k=4$ — — — — Z_3

Z_1

Z_4 S_4

$R_2 = R_4$ No $R_1 := R_1 + R_3$ ◄3rd stage $k=4$-- Z_5

Yes

STOP

Z_3 S_3

$R_3 = R_1$ No

Yes

Z_5 S_1

S_5

$R_3 = R_5$ No

Yes

$I_0: Z_3$

$I_1: J_{3,1,4}$

$I_2: S_3$

$I_3: J_{1,1,1}$

$R_3 := R_1$ with $k = 4$

$I_4: Z_1$

$I_5: Z_4$

$I_6: J_{2,4,14}$

$I_7: Z_5$

$I_8: J_{3,5,12}$

$I_9: S_5$

$I_{10}: S_1$

$I_{11}: J_{1,1,8}$

$I_{12}: S_4$

$I_{13}: J_{1,1,6}$

$I_{14}: STOP$

$R_1 := R_1 + R_3$, with $k = 4$.

N.B. k has not changed *only* because $R_3 := R_1$ uses no new registers. 7 has been added to the jump in I_8 and I_{11} as the subprogram starts at I_7, *not* I_0.

We note a final precaution which it is necessary to observe if programs are to be usable as subprograms. All registers named in the program, other than the answer register, must remain unchanged *at the end*. Notice that R_j has its original value at the end of $R_i: = R_i + R_j$. The reader can see why this restriction is necessary by going round the relevant loop in $R_i: = R_i \cdot R_j$ above.

EXERCISE 8.6.   Using any previously given program which you need, devise a flow diagram, $R_i: = R_i^{R_j}$, where $i \neq j$, to compute the power function a^b. (Hint: $a^0 = 1$, $a^{n+1} = a^n \cdot a$.) ∎

EXERCISE 8.7.   Use your flow diagram for $R_1: = R_1^{R_2}$ and expand in stages any subprograms, as was done for $R_1: = R_1 \cdot R_2$ above. ∎

An algorithm for writing programs from flow diagrams

Since translating the (full) flow diagram into a program gets increasingly cumbersome, we conclude our remarks on flow diagrams with an algorithm for writing a program from a flow diagram.

(1)   Associate with each instruction box ☐ and decision box ◇ a unique letter (its 'label'). Start with the first instruction to be obeyed, and follow the arrows. Follow round a loop before going on, and finish with STOP. Strictly speaking, the order of labelling does not matter, but it is less 'messy' to proceed in this order.

(2)   Write, in a vertical list, all the labels in order of labelling: to the right of each label write its associated instruction and to the right of each instruction write the label of the next box encountered by following the arrows. In the case of a decision box labelled

write A: $J_{i,j,E}$: B.

(3)   If, in this list, the label, L, on the right of line t differs from the label on the left of the line t + 1 insert between the two $J_{1,1,L}$.

(4)   Number the lines in the list consecutively from 0.

(5)   Replace the label in each jump instruction with the number of the instruction so labelled.

(6)   Erase all labels.

The reader might like to use this algorithm on previous flow diagrams to see how it works; alternatively, try the next exercise.

EXERCISE 8.8 Use this algorithm on the answer to Exercise 8.7 and write out the full program for $R_1 := R_1^{R_2}$. ■

We know that any flow diagram which computes a function yields a program for that function. In future, given a flow diagram, we shall speak freely of 'the program' which computes the same function.

R-computability defined

Before we can define precisely when a function is R-computable we need some preliminaries.

Let $f_n(x_1, \ldots, x_n)$ be a numerical function of n variables and let P be a program. We associate with P a function of n variables, $P_n(x_1, \ldots, x_n)$ whose *value* for the n-tuple of natural numbers $a_1, \ldots, a_n$ is defined as follows.

DEFINITION 8.1. Let $a_1, \ldots, a_n$ be stored in registers $R_1, \ldots, R_n$ respectively, all other registers being empty. Starting with 0 in the instruction counter, R_0, let the R-machine carry out the program P. If, after a finite number of moments, the machine reaches the instruction STOP then $P_n(a_1, \ldots, a_n)$ is defined and its *value* is taken to be the number which is stored at that moment in R_1. If STOP is never reached $P_n(a_1, \ldots, a_n)$ is undefined. ■

DEFINITION 8.2. A numerical function $f_n(x_1, \ldots, x_n)$ is said to be *computed by the program* P, and P is said to be a *program for computing* $f_n(x_1, \ldots, x_n)$, if, for all n-tuples of natural numbers $a_1, \ldots, a_n$ *either* both $f_n(a_1, \ldots, a_n)$ and $P_n(a_1, \ldots, a_n)$ are undefined, *or* both are defined and have the same value, i.e. $f_n(a_1, \ldots, a_n) = P_n(a_1, \ldots, a_n)$. ■

These preliminaries enable us to define R-computability.

DEFINITION 8.3. A function is *R-computable* if it is computed by some program. ■

EXERCISE 8.9. How can you decide if a function is R-computable? ■

Since a numerical function computed by a program may be undefined for some of its arguments and since the reader may not have encountered such functions before, we should say a little more about them.

DEFINITION 8.4. If a function is undefined for one or more of its arguments, we call that function *partial*. Otherwise, the function is said to be *total*. ∎

The predecessor function, which for each natural number n yields the number $n - 1$, is a partial function, since it is undefined at 0. The successor function is, of course, total.

We also note that partial functions $f(x_1,\ldots,x_n)$ and $g(x_1,\ldots,x_n)$ are equal whenever, for all n-tuples $a_1,\ldots,a_n$, either f and g are both undefined, or f and g are both defined and

$$f(a_1,\ldots,a_n) = g(a_1,\ldots,a_n).$$

Church's Thesis

The obvious question to ask about R-computability is, 'How general is it? Are there are any intuitively computable functions which are not R-computable?' There is overwhelming evidence for what has come to be known as Church's Thesis — that all (intuitively) computable functions are R-computable. In 1936 Alonzo Church (1936a) gave various arguments in its favour. Of course Church's Thesis cannot be rigorously proved since it refers to an informal notion, (intuitive) computability. However, evidence for it has accumulated since its formulation.

There are two main kinds of evidence. First, as we shall see in the next chapter, many computable functions can be shown to be R-computable and mathematicians generally agree that attempts to construct intuitively computable functions which are not R-computable have all failed. (This is also true of diagonalization, as we shall see.) Second, as we mentioned earlier, numerous, *independent* attempts at defining the computable functions have produced *equivalent* definitions. That diverse definitions, in terms of λ-definability, recursiveness, Turing machines, and binormality should all turn out to be equivalent is widely regarded as establishing that some 'absolute' notion of computability has been defined.

After the reader has worked through the next chapter he will probably be convinced of the truth of Church's Thesis, but it might be instructive to try to construct an intuitively computable function which is not R-computable. The converse of Church's Thesis, that every R-computable function is computable, can hardly be doubted, so we shall proceed to identify the R-computable and computable functions in what follows.

9
PROGRAMMING COMPUTATIONS

This chapter has several purposes. In the first place we show that various number-theoretic functions are R-computable. This partly fulfils the promise of the previous chapter and also prepares the ground for the next part, where we give a flow diagram to show that 'v is the gödel number of a term' is R-computable. Since it is clearly a mechanical matter to decide whether an expression is a term of N, say, this further realizes our previous promise. After this, it should require only a little further argument to convince the reader that an R-machine can decide for any v whether v is the gödel number of a formula. The reader should then find it easy to believe that 'v is the gödel number of a formula with exactly one free variable', 'v is the gödel number of a proof', and related metamathematical properties are R-computable.

Examples of R-computable functions

Bearing in mind that we may use any program given previously, we now give a flow diagram to show that $n!$ (n factorial) is R-computable.

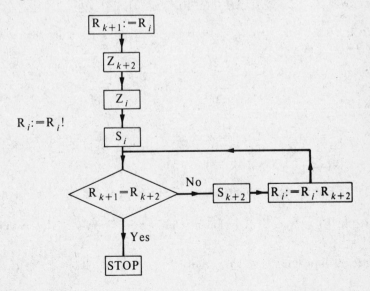

EXERCISE 9.1.   How do you check that the flow diagram is correct? ■

We give some further examples as exercises. In each devise a flow diagram to compute the given function. Remember to express each diagram in a form suitable for use as a subprogram by using generalized registers R_i and R_j, and by leaving all registers named in the function other than the answer register, unchanged at the end. Remember also that in answering a question any previously given program may be used as a subprogram.

EXERCISE 9.2.

$$R_i := \delta(R_i) = \begin{cases} r_i - 1 & \text{if} \quad r_i \neq 0 \\ 0 & \text{if} \quad r_i = 0 \end{cases} = \text{the predecessor of } r_i.$$

(Hint: $R_{k+1} := R_i$, Z_i, then $n = \delta(R_{k+1})$ if both $n = 0$ and $r_{k+1} = 0$, or if $n + 1 = r_{k+1}$.) ■

EXERCISE 9.3.

$$R_i := R_i \dotminus R_j = \begin{cases} r_i - r_j & \text{if} \quad r_i \geqslant r_j \\ 0 & \text{if} \quad r_i < r_j \end{cases} \qquad \text{Assume } i \neq j.$$

(Hint: use the δ-function on both R_i and R_j.) ■

EXERCISE 9.4. $R_i := \max(R_i, R_j) = $ the maximum of r_i and r_j. Assume $i \neq j$.

$$\left(\text{Hint:} \quad \max(R_i, R_j) = \begin{cases} r_i & \text{if} \quad r_j \dotminus r_i = 0. \\ r_j & \text{if} \quad r_j \dotminus r_i \neq 0. \end{cases} \right) \quad ■$$

EXERCISE 9.5. $R_i := \mathrm{rm}(R_i, R_j) = $ the remainder of r_i upon division by r_j. Assume $r_j \neq 0$ and $i \neq j$.

(Hint: you might use $R_i := R_i \dotminus R_j$ repeatedly until $r_i = r_j$ or $r_i < r_j$). ■

EXERCISE 9.6.

$$R_i := \mathrm{qt}(R_i, R_j) = \text{the largest integer} \leqslant \frac{r_i}{r_j}, \text{ or the quotient of } \frac{r_i}{r_j}.$$

Assume $r_j \neq 0$ and $j \neq j$.

(Hint: as in Exercise 9.5, count the number of times you can subtract r_j from r_i). ■

EXERCISE 9.7. $R_i := D(R_j) = $ the number of positive divisors of r_j. Assume $i \neq j$ and that $D(0) = 1$.

(Hint: for all n such that $1 \leqslant n \leqslant r_j$, n divides r_j if and only if $0 = \mathrm{rm}(r_j, n)$.) ■

With the aid of the foregoing programs we can show that the predicate 'n is a prime number' is R-computable. But first we give two definitions.

DEFINITION 9.1.   Let $a_1, \ldots, a_n$ be an n-tuple of numbers and let $P(x_1, \ldots, x_n)$ be a number-theoretic predicate. The *characteristic function of* P, C_p is defined as follows:

$$C_p(a_1, \ldots, a_n) = \begin{cases} 0 & \text{if } P(a_1, \ldots, a_n) \text{ is true} \\ \\ 1 & \text{if } P(a_1, \ldots, a_n) \text{ is false.} \end{cases} \blacksquare$$

DEFINITION 9.2.   A *predicate* is said to be *R-computable* if and only if its characteristic function is R-computable. $\blacksquare$

EXERCISE 9.8.   Devise a flow diagram to compute the function

$$R_i := Pr(R_j) = \begin{cases} 0 & \text{if } R_j \text{ is prime} \\ \\ 1 & \text{otherwise.} \end{cases}$$

Assume that $i \neq j$.

(Hint: $0 = Pr(R_j)$ if $D(R_j) = 2$.) $\blacksquare$

Computability of formal notions

In the previous examples we have seen that various familiar functions are R-computable. Although they were obviously intuitively computable it took some work to show that the R-machine — with specific and restricted sets of instructions — could compute each of them. It is also obvious that, given our gödel numbering from Chapter 4, the predicate 'v is a gödel number of a term' is intuitively computable. We need only write out the prime factorization of v, convert the powers into symbols, and check the resulting expression (cf. Exercise 4.8). But it is not so obvious that an R-machine can do this for any v. Suppose that v has as prime factorization

$$v| = p_0^{a_0} \cdot \ldots \cdot p_n^{a_n}$$

where some, but not all, of the a_is may be zero. It might be thought that we can find a program P_t which proceeds broadly as follows: given v in R_1, P_t writes $a_0, \ldots, a_n$ successively in registers $R_2, \ldots, R_{n+2}$. P_t then reads registers $R_2, \ldots, R_{n+2}$ to check whether v is the gödel number of a term, and writes the answer in R_1.

EXERCISE 9.9.   Show that there cannot be such a program P_t.
(Hint: P_t would have to work *for any* v, however large.) $\blacksquare$

The R-machine cannot 'display' the prime factorization of v in the way just described. Instead, we can program it to proceed broadly as follows: if v is in R_j, we attempt to 'reconstruct' v In R_{k+1} by using the rules for defining a term. To do this, we shall need to be able to read the 'information' contained in v — we must be able to inspect its prime factorization. In particular, we shall need to be able to compute the jth prime, p_j, and its power a_j in the prime factorization of v. To see how this would go, the reader should try devising flow diagrams for the following functions.

EXERCISE 9.10. $R_i := (R_j)_{R_q} = n$, where $r_q^n | r_j$ and $r_q^{n+1} \!\!\not|\, r_j$. Assume r_j and $r_q \neq 0$, and $i \neq j \neq q$.

(Hint: If $r_q^n | r_j$, then $r_q^{n+1} | r_j$ iff $r_q | \mathrm{qt}(R_j, R_q^n)$.) ∎

EXERCISE 9.11. $R_i := p_{R_j} =$ the (r_j)th prime. Assume $i \neq j$. Note that $p_0 = 2$.

(Hint: starting from $p_0 = 2$, you can look for successively larger primes using 9.8.) ∎

EXERCISE 9.12. $R_i := (R_j)_{p_{R_q}} = n$, where $p_{R_q}^n | r_j$ and $p_{R_q}^{n+1} \!\!\not|\, r_j$. Assume $i \neq j \neq q$.

(Hint: very simple, using 9.10 and 9.11.) ∎

Furthermore, in reading the information contained in v, we shall need to be able to compute:

(A)   the k such that p_k is the *greatest* prime divisor of v;

(B)   the k such that p_k is the *smallest* prime divisor of v;

(C)   for a given prime p_j, that g such that p_g is the *least* prime *greater* than p_j which does not divide v; and

(D)   for a given prime p_j, that s such that p_s is the *greatest* prime *smaller* than p_j which does not divide v.

We conclude this section with suitable flow diagrams for (A), (C), and (D). The reader should check that they are correct. (B) is obvious, given (A).

(A) If $i \neq j$ and $r_j \geqslant 2$, NGPD(R_j) = the number of the greatest prime divisor of r_j, i.e. the largest n such that $p_n | r_j$.

$R_i := \text{NGPD}(R_j)$

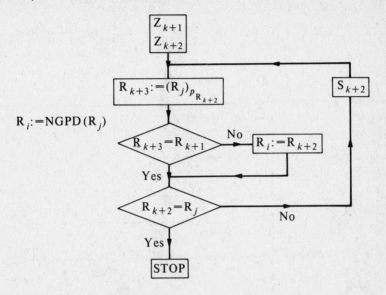

(C) Assuming that $i \neq j \neq m$, LG(R_j, R_m) = least n such that $n > r_m$ and $(R_j)_{p_n} = 0$. (We use Exercise 9.11 on successively larger $n > r_m$.)

$R_i := \text{LG}(R_j, R_m)$

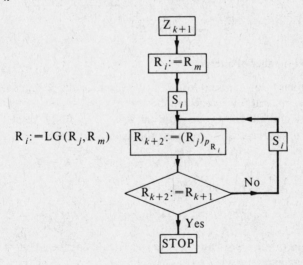

(D) Assuming $i \neq j \neq m$,

$$GL(R_j, R_m) = \begin{cases} 0 \text{ if } r_m = 0 \text{ or if } (R_j)_{p_n} \neq 0 \text{ for all } n < r_m \\ \\ \text{greatest } n \text{ such that } n < r_m \text{ and } (R_j)_{p_n} = 0, \text{ otherwise.} \end{cases}$$

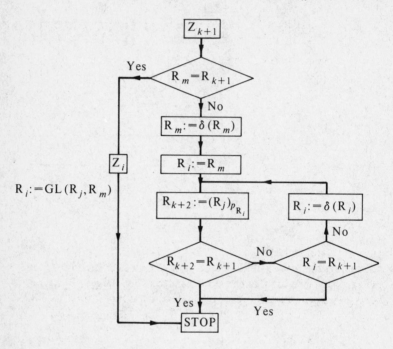

v is the gödel number of a term

We now outline a flow diagram for testing whether v is the gödel number of a term. We assume that some gödel numbering has already been written into the program; any one will do of course, but the reader may find it helpful to bear in mind the one we gave earlier. We define the operation 'Trm' as follows

$$R_i := \mathrm{Trm}(R_j) = \begin{cases} 0 \text{ if } R_j \text{ is the gödel number of a } term; \\ \\ 1 \text{ otherwise.} \end{cases}$$

Remember that $r_j = p_m^{x_m} \cdot \ldots \cdot p_n^{x_n}$, so that $x_i = (R_j)_{p_i}$. Let $r_{k+1} = p_m^{y_m} \cdot \ldots \cdot p_n^{y_n}$; hence $y_i = (R_{k+1})_{p_i}$.

In the light of the preceding examples, we should now be able to abbreviate our flow diagrams. Abbreviations will be surrounded by broken lines, for example,

We shall write $\underline{var}, \underline{0}, \underline{+}, \underline{\cdot}, \underline{'}, \underline{(}, \underline{)}$ for the gödel numbers of a variable, or of the symbol 0, etc.

$$\diamond x_{r_{k+3}} = \underline{var} \text{ or } \underline{0} \diamond$$

will be read 'Is $x_{r_{k+3}}$ the gödel number of a variable or of the symbol 0?'

$$\diamond y_m = 0 \diamond$$

will be read 'Is $y_m = 0$?', *not* 'Is y_m the gödel number of the symbol 0?'

Finally,

$$\diamond R_{k+1} = n \diamond$$

will be read 'Does R_{k+1} contain the number n?'.

There are many ways of proceeding. Broadly speaking we shall attempt to reconstruct R_j in R_{k+1} by the rules for defining a term. The rules state that 0 is a term, a variable is a term, and if t and s are terms, so are t', $(t + s)$ and $(t \cdot s)$. The program is rather long, so we identify five different parts of it to make it easier to follow.

(I) We first check that all the x_is, where $m \leqslant i \leqslant n$, are strictly positive. Otherwise, v is not the gödel number of a term. We put any occurrences in v of variables or of 0 into R_{k+1}.

(II) We check that, if $q < u$ and x_q and x_u are the gödel numbers of 0 or of variables, there is an x_w such that $q < w < u$ and x_w is the gödel number of $\cdot$ or $+$.

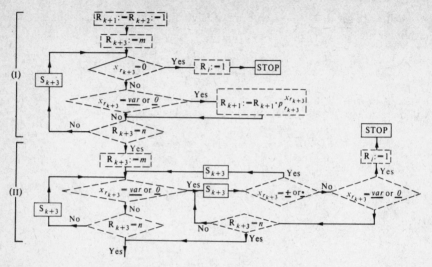

We also put parts III, IV, and V together.

(III) We successor in R_{k+1} any symbol which has *already* been put into R_{k+1} and is successored in R_j. Note that if R_{k+3} contains the number z, then $r_{k+3} + 1 = z + 1$ and $r_{k+3} \div 1 = z \div 1$.

(IV) We now put into R_{k+1} those occurrences of $(t + s)$ and $(t \cdot s)$ in R_j whose terms, t and s, have *already been reconstructed* in R_{k+1}.

(V) Finally, we ask if $R_{k+1} = R_j$. If not, we check to see if going again through programs III and IV would change R_{k+1}.

The reader should check that this flow diagram is correct.

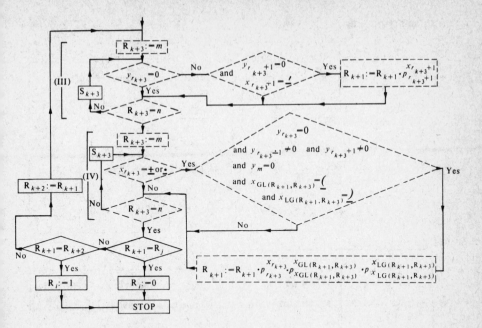

We should say a word about how

$$x_{GL(R_{k+1},R_{k+3})} = \underline{(}$$

is to be read. We already know that $x_{r_{k+3}} = \underline{+}$ or $\underline{\cdot}$. Let $GL(R_{k+1}, R_{k+3}) = g$, where $GL(R_{k+1}, R_{k+3}) =$ the greatest $n < r_{k+3}$ such that $p_n \nmid r_{k+1}$, or is 0 if there is no such n. Then, (A) asks whether v contains '(' where it should if we are about to reconstruct a term $(t + s)$ or $(t \cdot s)$, i.e. whether $x_g = \underline{(}$. Similar remarks apply to

$$x_{LG(R_{k+1},R_{k+3})} = \underline{)}.$$

A summary outline

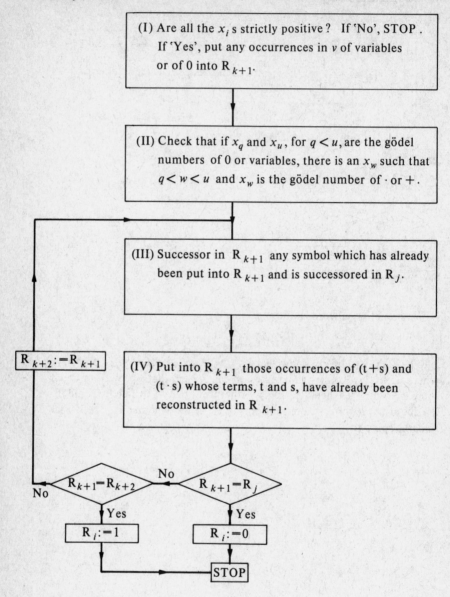

(I) Are all the x_i s strictly positive ? If 'No', STOP .
    If 'Yes', put any occurrences in v of variables
    or of 0 into R_{k+1}.

(II) Check that if x_q and x_u, for $q < u$, are the gödel
    numbers of 0 or variables, there is an x_w such that
    $q < w < u$ and x_w is the gödel number of · or $+$.

(III) Successor in R_{k+1} any symbol which has already
    been put into R_{k+1} and is successored in R_j.

$R_{k+2} := R_{k+1}$

(IV) Put into R_{k+1} those occurrences of (t$+$s) and
    (t · s) whose terms, t and s, have already been
    reconstructed in R_{k+1}.

$R_{k+1} = R_{k+2}$ No $R_{k+1} = R_j$

No

Yes   Yes

$R_i := 1$ $R_i := 0$

STOP

Checking the flow diagram

We give the full flow diagram below for ease of reference.

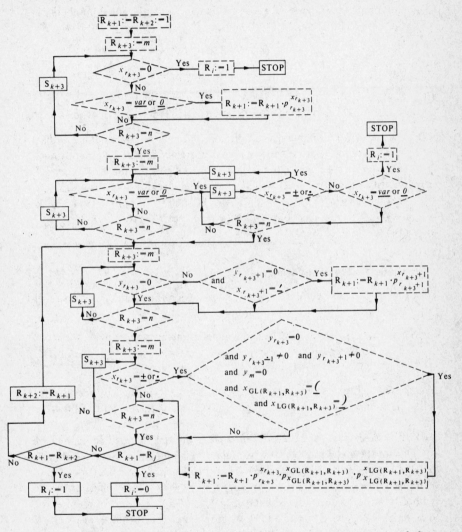

The reader should do as many parts of the following exercise as are needed to convince him or her that the flow diagram does what was intended.

EXERCISE 9.13.

(i)   How do we know that the R-machine can obey $\lceil R_{k+3} := m \rceil$?

(ii) Why, at stage (III), do we need to ask if $y_{r_{k+3}+1} = 0$?

(Hint: consider $(a' + 0)'$.)

(iii) Determine the order in which the R-machine puts the gödel numbers of the symbols which make up the term

$$(a + (0 + (a \cdot b')')')$$

into R_{k+1}. Assume that the number of this term is already in R_j.

(iv) Say which questions in stage (IV) establish that the following are *not* terms: (1) $((a \cdot b))$; (2) $a(+ b)$; (3) $(a + (b)$; (4) $(a + b) + c)$, when $m = 0$.

(v) For each of the following examples, assume that its gödel number, v, is in R_j and check that our flow diagram correctly decides whether or not it is a term. In each case in which v is *not* the gödel number of a *term*, identify the point in the flow diagram which decides that it is not.
(1) a0a; (2) a″a; (3) ′+′; (4) 0; (5) a; (6) (0); (7) a‴; (8) a·b; (9) (a + b);
(10) (′a + b); (11) a + b + c; (12) (a + b) + c); (13) ((a·0) + (b·0));
(14) (a·((a + b)′ + c)′); (15) (a + b)a. ∎

v is the gödel number of a formula

The program for testing if v is the gödel number of a formula is closely parallel to the one for $R_i := \mathrm{Trm}(R_j)$. For formulas,

$$R_i := \mathrm{Form}(R_j) = \begin{cases} 0 \text{ if } R_j \text{ is the gödel number of a formula} \\ \\ 1 \text{ otherwise} \end{cases}$$

Again, $R_j = p_m^{x_m} \cdot \ldots \cdot p_n^{x_n}$, and we try to reconstruct R_j in R_{k+1} by the rules defining *formula*. We outline the program now, leaving the reader to fill in the details.

We first check that all the x_is, for $m \leqslant i \leqslant n$, are strictly positive. If they are not, v is not the gödel number of a formula, and we set $R_i := 1$ and STOP. If they are, we put any occurrences of '=' in v into R_{k+1}.

We now check that if x_q and x_u with $q < u$ are the gödel numbers of the symbol '=', there is an x_w such that $q < w < u$ and x_w is the gödel number of ∨, &, or ⊃. If not, $R_i := 1$ and STOP. What remains is the problem of reconstructing R_j in R_{k+1} by the rules defining *formula*. Recall that, if t, s are terms then (t = s) is a formula. if A, B are formulas, so are (A ⊃ B), (A & B), (A ∨ B), and ¬A, and, if x is a variable and A is a formula, then ∀xA and ∃xA are formulas. There are no other formulas. The only 'complication' is in the formation of prime formulas, those formulas of the form (t = s) where t, s are terms. To reconstruct these in R_{k+1}, we identify the leftmost occurrence of '=' in R_{k+1} and then use $\mathrm{Trm}(R_j)$ on successively longer sequences to the left and to the right of '='. If this does not yield a prime formula (t = s), we set $R_i := 1$ and STOP. If it does, we proceed to the next '=' to the right and repeat the process.

Extrapolation

Although programs rapidly get more complicated, it should by now be possible for the reader to believe that there are programs which, even given the restricted instructions available to the R-machine, can decide whether v is the gödel number of a formula with exactly one free variable or whether v is the gödel number of a proof.

For a detailed proof of this see, for example, Mendelson (1964). Mendelson uses recursive functions, which are equivalent to R-computable functions.

10
NUMERALWISE REPRESENTATION

We showed earlier (Chapter 6) that formal analogues of certain informal number-theoretic results were provable in **N**. Even without *formal* symbols, '$<$', '$|$', and 'Pr( )', we were able by means of suitable definitions to find formulas of **N** which enabled us to prove in **N** recognizable formal analogues of some of the usual basic properties of $<$, $|$, and Pr( ).

The properties we considered there were all general properties, like the transivity of $<$, or Euclid's theorem. We did not show that if, for natural number numbers m and n, $m < n$, then the formal analogue for the corresponding numerals is provable in **N**, i.e. $\vdash_N m < n$. Nor did we show that if $m \nless n$, then $\vdash_N \neg m < n$. But we should clearly expect these to be the case if **N** is to have any claim to adequacy for number theory. We should expect the same for other relations. Clearly, we would expect that if $m | n$ then $\vdash m | n$ and if $m \nmid n$ then $\vdash \neg m | n$. For numbers k, m, and n and the relations $m \equiv n(\bmod\ k)$ or $\mathrm{rm}(m, n) = k$, we would expect to find formulas of **N**, say $P_\equiv(x_1, x_2, x_3)$ and $P_{\mathrm{rm}}(x_1, x_2, x_3)$, such that x_1, x_2, and x_3 are distinct variables and

$$\text{if} \quad m \equiv n(\bmod\ k) \quad \text{then} \quad \vdash P_\equiv(m, n, k), \quad \text{while}$$

$$\text{if} \quad m \neq n(\bmod\ k) \quad \text{then} \quad \vdash \neg P_\equiv(m, n, k); \quad \text{and}$$

$$\text{if} \quad \mathrm{rm}(m, n) = k \quad \text{then} \quad \vdash P_{\mathrm{rm}}(m, n, k), \quad \text{while}$$

$$\text{if} \quad \mathrm{rm}(m, n) \neq k \quad \text{then} \quad \vdash \neg P_{\mathrm{rm}}(m, n, k).$$

Numeralwise expressibility and representability

We cast these intuitions into the form of a definition.

DEFINITION 10.1. An informal number-theoretic predicate $P(x_1,\ldots,x_n)$ is said to be *numeralwise expressible* in a formal system (in our case, **N**), if there is a formula $P(x_1,\ldots,x_n)$ with exactly the n variables $x_1,\ldots,x_n$, such that for each particular n-tuple $m_1,\ldots,m_n$ of numbers,

$$\text{if } P(m_1,\ldots,m_n) \text{ is true, then } \vdash P(m_1,\ldots,m_n) \text{ and}$$

$$\text{if } P(m_1,\ldots,m_n) \text{ is false, then } \vdash \neg P(m_1,\ldots,m_n). \blacksquare$$

EXERCISE 10.1. Prove that the relations (*164) $a = b$ and (*165) $a < b$ are numeralwise expressed by the formulas a = b and a < b, respectively. Recall that a < b abbreviates $\exists c(c' + a = b)$. ∎

We also noted earlier (Chapter 6) that we have no obvious way of representing the power function, a^b, or a factorial, $a!$, in **N**. **N** is clearly inadequate even for the elementary number theory expounded initially in Chapter 1, unless these functions are in some suitable sense representable in **N**. What then do we want?

Consider a^b. At the very least we want some formula $A_p(x_1, x_2, x_3)$ of **N** such that x_1, x_2, x_3 are distinct variables and

$$\text{if } m^n = k \text{ then } \vdash A_p(m, n, k).$$

But the essential thing about a function is that it has a unique value for each argument; we want to be sure the formula $A_p(x_1, x_2, x_3)$ reflects that characteristic: that for arbitrary arguments m, n it is formally provable that

$$\exists! x_3 A_p(m, n, x_3).$$

$\exists! xA(x)$ is read 'There exists a unique x such that $A(x)$' and is an abbreviation for

$$\exists x[A(x) \mathbin{\&} \forall y(A(y) \supset x = y)]$$

where y is distinct from x and does not occur free in $A(x)$ but is free for x in $A(x)$. If we can find a formula A_p which satisfies both the above conditions, then many interesting properties of powers become provable in **N**.

EXERCISE 10.2. What do we need in order to represent $a!$ in **N**? ∎

DEFINITION 10.2. An informal (total) number-theoretic function $f(x_1,\ldots,x_n)$ is *numeralwise representable* in a formal system (in our case, **N**) if there is a formula $F(x_1,\ldots, x_n, y)$ with exactly the $n + 1$ variables $x_1,\ldots, x_n, y$ such that for each particular n-tuple $m_1,\ldots, m_n$ of numbers

(i)   if $f(m_1,\ldots, m_n) = m_{n+1}$, then $\vdash F(m_1,\ldots, m_n, m_{n+1})$,

and

(ii) $\vdash \exists! y F(m_1,\ldots, m_n, y)$. ∎

EXERCISE 10.3. Show that if the function $f(x_1,\ldots, x_n)$ is *numeralwise represented* by the formula $F(x_1,\ldots, x_n, y)$, this formula *numeralwise expresses* the predicate $f(x_1,\ldots, x_n) = y$, the 'graph' of f.
(Hint: Assume $a = b$ is numeralwise expressible (*164) and that $r \neq s$, $A(r)$, $\exists! xA(x) \vdash \neg A(x)$ (*173).) ∎

EXERCISE 10.4.   Show that the functions

$$(*175)\, a', \quad (*176)\, a + b, \quad (*177)\, a \cdot b$$

are numeralwise represented by the formulas $a' = b$, $a + b = c$, and $a \cdot b = c$ respectively.

(Hint: if t is a term not containing x, $\vdash \exists! x(t = x)$ ($*171$).) ∎

We are now in a position to prove ($*180$), that Gödel's β-function (Chapter 1, page 11):

$$\beta(c, d, i) = \mathrm{rm}(c, (i' \cdot d)')$$

is numeralwise representable in **N**.

THEOREM 10.1.   The β-function is numeralwise represented in **N** by the formula B(c, d, i, w):

$$\exists u(c = (i' \cdot d)' \cdot u + w \;\&\; w < (i' \cdot d)').$$

Proof. By $*135$(b), $\vdash 0 < (i' \cdot d)'$. So, by $*146$(a), (b), and *modus ponens*,

$$\vdash \exists! w \exists u(c = (i' \cdot d)'u + w \;\&\; w < (i' \cdot d)').$$

Hence, for numbers k, m, n, we have

(ii) $\vdash \exists! w \exists u(k = (n', m)' \cdot u + w \;\&\; w < (n' \cdot m)').$

To prove condition (i), assume $\beta(k, m, n) = r$. Then, there is a number q such that

$$k = (n' \cdot m)' \cdot q + r \quad \text{and} \quad r < (n' \cdot m)'.$$

Hence, by $*175$, $*176$, $*177$, and $*164$, $*165$ (Exercises 10.1 and 10.4),

$$\vdash k = (n' \cdot m)' \cdot q + r \quad \text{and} \quad \vdash r < (n' \cdot m)'.$$

By elementary logic, it follows immediately that

(i) $\vdash \exists u(k = (n' \cdot m)' \cdot u + r \;\&\; r < (n' \cdot m)').$ ∎

Notice that condition (ii) in the definition of numeralwise representability is weaker than the condition

(ii)′ $\vdash \exists! x F(a, b, c, \ldots, x)$

where a, b, c, and x are *variables*. The corresponding notion, defined by conditions (i) and (ii)′, is called *strong representability*. It is easy to see that a', $a + b$, $a \cdot b$ and $\beta(c, d, i)$ are *strongly* representable in **N**.

EXERCISE 10.5.   How does one see this? ∎

The numeralwise representability of R-computable functions

We now have what we need to prove that all total R-computable functions are numeralwise representable in **N**. Although the proof of this result is straightforward, some of the details are rather involved. To help the reader distinguish the wood from the trees, we first explain the idea behind the proof and then present the proof itself. Many of the details appear as exercises.

The idea of the proof is this. If $f(x_1,\dots,x_n)$ is an R-computable function, then some program P_{h+1}, with $h + 1$ instructions, computes it. To each instruction I_j, for $j \leqslant h$, in that program, we use the β-function to assign a predicate which 'describes the action' of I_j. *Every such predicate is numeralwise expressible in* **N**, because the β-function is. The conjunction of all these predicates then 'describes the action' of the R-machine with program P_{h+1}, *and is also expressible in* **N**. Combining this formal expression with one expressing the 'start' condition of the R-machine, we obtain a formula $F_{P_{h+1}}(x_1,\dots,x_n,y)$ with exactly $n + 1$ free variables $x_1,\dots,x_n, y$ and which can be proved to satisfy the representability conditions for the function $f(x_1,\dots,x_n)$.

THEOREM 10.2. All total R-computable functions are numeralwise representable in **N**.

Proof. Let $f(x_1,\dots,x_n)$ be total and R-computable by program P_{h+1}, with instructions $I_0,\dots, I_h$. Then we must find a formula $F_{P_{h+1}}$, with exactly $n + 1$ free variables, such that for numbers $k_1,\dots, k_{n+1}$

$$\text{if } f(k_1,\dots,k_n) = k_{n+1} \text{ then } \vdash_N F_{P_{h+1}}(k_1,\dots,k_n, y) \sim y = k_{n+1}.$$

(This is equivalent to representability conditions (i) and (ii). See Exercises 10.10 and 10.11.)

Let $m =$ the maximum i such that either $i = n$ or R_i is referred to in P_{h+1} and let t^* be a moment at which P_{h+1} halts. Then, we use,

$$\beta(c_0, d_0, t) \text{ for } t = 0, 1,\dots, t^*$$

to represent the number of the instruction about to be carried out at moment t, and

$$\beta(c_i, d_i, t) \text{ for } 1 \leqslant i \leqslant m \text{ and } t \leqslant t^*$$

to represent the contents of R_i at moment t.

EXERCISE 10.6. The program given on p. 92 for computing $R_1 + R_2$ is as follows:

$$I_0 = Z_3$$
$$I_1 = J_{2,3,5}$$
$$I_2 = S_3$$
$$I_3 = S_1$$
$$I_4 = J_{1,1,1}$$
$$I_5 = \text{STOP}.$$

(i) For this program what are the values of n, h, and m?

(ii) (a) Write out, in tabular form, the computation by this program of $3 + 2$ starting:

Moment	Instruction about to be obeyed	Contents of R_1	R_2	R_3	...
0	0	3	2	0	
1	1	...	...	...	

(b) For this example what are the values of

$$t^*, \beta(c_0, d_0, t^*), \beta(c_1, d_1, 7)? \quad \blacksquare$$

To each instruction I_j we assign a predicate describing its action at moment t:

If I_j is S_k, then assign to I_j, where $1 \leqslant i \leqslant m$ and $i \neq k$,

$$\beta(c_0, d_0, t) = j \supset \{\beta(c_0, d_0, t+1) = j+1$$
$$\& \ \beta(c_k, d_k, t+1) = 1 + \beta(c_k, d_k, t)$$
$$\& \ \forall i(\beta(c_i, d_i, t+1) = \beta(c_i, d_i, t))\}.$$

If I_j is Z_k, then assign to I_j, where $1 \leqslant i \leqslant m$ and $i \neq k$,

$$\beta(c_0, d_0, t) = j \supset \{\beta(c_0, d_0, t+1) = j+1$$
$$\& \ \beta(c_k, d_k, t+1) = 0$$
$$\& \ \forall i(\beta(c_i, d_i, t+1) = \beta(c_i, d_i, t))\}.$$

If I_j is $J_{k,r,q}$, assign to I_j, for $1 \leqslant i \leqslant m$,

$$\beta(c_0, d_0, t) = j \supset \{(\beta(c_k, d_k, t) = \beta(c_r, d_r, t) \supset \beta(c_0, d_0, t+1) = q)$$

$$\& \ (\beta(c_k, d_k, t) \neq \beta(c_r, d_r, t) \supset \beta(c_0, d_0, t+1) = j+1)$$

$$\& \ \forall i(\beta(c_i, d_i, t+1) = \beta(c_i, d_i, t))\}.$$

To I_h, which is STOP, assign

$$\beta(c_0, d_0, t^*) = h \ \& \ \beta(c_1, d_1, t^*) = y.$$

We also assign to the 'start' condition the following predicate which describes the initial state of the R-machine:

$$\beta(c_0, d_0, 0) = 0 \ \& \ \forall i(\beta(c_i, d_i, 0) = x_i) \ \& \ (\forall j \beta(c_j, d_j, 0) = 0).$$

for $1 \leqslant i \leqslant n$ and $n+1 \leqslant j \leqslant m$.

EXERCISE 10.7.   For the program of Exercise 10.6, write out with *numbers* inserted where possible, the predicates associated with (i) I_2, (ii) I_1, and (iii) I_5. ■

Now, as the β-function is representable, it follows that each of the $h+2$ predicates we have just assigned is expressible. For example,
$\beta(x, y, z) = 1 + \beta(x_1, y_1, z_1)$ is expressed by

$$\forall w \forall w_1 [B(x, y, z, w) \ \& \ B(x_1, y_1, z_1, w_1) \supset w = w_1 + 1]$$

EXERCISE 10.8. (Hard)   Show that $\beta(x, y, z) = 1 + \beta(x_1, y_1, z_1)$ is numeralwise expressible in **N** by the formula

$$\forall w \forall w_1 [B(x, y, z, w) \ \& \ B(x_1, y_1, z_1, w_1) \supset w = w_1 + 1]. \ ■$$

For each j, $0 \leqslant j \leqslant h$, let $A_j(t)$ be the *formal* predicate expressing the informal predicate assigned to I_j. Let $H(t^*, y)$ express the predicate of I_h and $S(x_1, \ldots, x_n)$ that of the 'start' condition. Define $A(t)$:

$$A(t) \sim A_0(t) \ \& \ \ldots \ \& \ A_{h-1}(t) \ \& \ \forall r B(c_0, d_0, t, r) \supset r < h].$$

Then let $F_{P_{h+1}}$ be the formal predicate, with free variables $x_1, \ldots, x_n, y$,

$$\exists t^* \exists c_0 \exists d_0 \ldots \exists c_m \exists d_m [S(x_1, \ldots, x_n) \ \& \ \forall t(t < t^* \supset A(t)) \ \& \ H(t^*, y)].$$

EXERCISE 10.9. $F_{P_{h+1}}$ is different for different programs. Describe briefly its features for the program on p. 169 for computing the power function a^b. ■

We now show that $F_{P_{h+1}}$ satisfies the representability conditions. First, it is easy to see that

(i)   if $f(k_1, \ldots, k_n) = k_{n+1}$ then $\vdash_N F_{P_{h+1}}(k_1, \ldots, k_n, k_{n+1}).$

EXERCISE 10.10.   Show that if $f(k_1,\ldots,k_n) = k_{n*1}$ then $\vdash_N F_{P_{h+1}}(k_1,\ldots,k_n,k_{n+1})$. ∎

It remains to show the uniqueness condition,

(ii) $\vdash_N \exists! y F_{P_{h+1}}(k_1,\ldots,k_n,y)$.

We do this by showing that

(ii′)   if $f(k_1,\ldots,k_n) = k_{n+1}$ then $\vdash_N F_{P_{h+1}}(k_1,\ldots,k_n,y) \supset y = k_{n+1}$.

EXERCISE 10.11.   Show that (ii′) does establish (ii). ∎

In preparation for the repeated use of Axiom 12, let Φ be $F_{P_{h+1}}(k_1,\ldots,k_n,y)$ *without* the initial quantifiers $\exists t^*, \ldots, \exists d_m$. We first show, by induction in the metalanguage, that if $t^*, c_0, d_0, \ldots, c_m, d_m$ are formal numbers corresponding to a genuine computation of $f(k_1,\ldots,k_n)$ by P_{h+1}, then for $t = 0, \ldots, t^*$, and for each $i, 0 \leqslant i \leqslant m$,

$$\vdash \Phi \supset \forall w \forall w_1 [B(c_i, d_i, t, w)\ \&\ B(c_i, d_i, t, w_1) \supset w = w_1].$$

Remember that the cs and ds are definitely not unique, since the β-representation is not. Also, we have yet to show that t^* is unique.

EXERCISE 10.12. (Hard)   Prove the preceding assertion.

(Hint: for $i = 0$, $\vdash \Phi \supset B(c_0, d_0, 0, 0)$ and $\vdash B(c_0, d_0, 0, 0)$; the representability of the β-function and the method of 10.8 easily yield the *basis*.) ∎

We conclude the proof in a number of simple steps. In particular, we have (i) $\vdash \Phi \supset B(c_0, d_0, t^*, h)$, since $\vdash B(c_0, d_0, t^*, h)$. Therefore, from (ii):

$$\vdash \Phi \supset \forall t(t < t^* \supset A(t)) \text{ and } \vdash A(t) \supset (B(c_0, d_0, t, r) \supset r < h),$$

we get, by elementary logic, (iii) $\vdash \Phi \supset \neg(t^* < t^*)$.
  But also, from $\vdash \Phi \supset H(t^*, y)$, we obtain (iv) $\vdash \Phi \supset B(c_0, d_0, t^*, h)$. Since (v):

$$\vdash t < t^* \supset (B(c_0, d_0, t, r) \supset r < h),$$

we can use Exercise 10.12 to get $\vdash \neg(t^* < t^*)$ and hence $\vdash \Phi \supset t^* = t^*$. Finally, by Exercise 10.12 again, (vi) $\vdash \Phi \supset B(c_1, d_1, t^*, k_{n+1})$ follows. Since $\vdash \Phi \supset B(c_1, d_1, t^*, y)$, we easily have (vii) $\vdash \Phi \supset y = k_{n+1}$. Repeated use of Axiom 12 then gives the required result. ∎

EXERCISE 10.13.   Give brief justifications for the preceding steps (i)–(vii). ∎

11

INCOMPLETENESS OF THE FORMAL THEORY

We began these notes with some results from elementary number theory. Because this body of theory is, by now, well established and well articulated, it was relatively easy to draw out the basic mathematical and logical results and assumptions, and to produce an axiomatic formal number theory, **N**. We convinced ourselves that **N** was adequate for the proofs of many of the fundamental results of number theory by proving them directly within **N**. However, many interesting questions *about* **N** remain unanswered.

Can we prove *all* the truths of number theory in **N**? In other words, is **N** *complete*?

Is **N** consistent, or can we prove contradictory results from the axioms we have chosen?

Can we decide for an arbitrary mathematical proposition, e.g. Fermat's last theorem, whether it is a theorem of **N**?

We are now in a position to answer these and some related questions.

The incompleteness of N

Gödel (1931) showed that formal number theory must be incomplete by actually producing an unprovable but true formula. Rosser (1936) modified this result, and it is his version of the result we now prove. We first define two predicates $\mathscr{P}$ (for 'Proof') and $\mathscr{R}$ (for 'Refutation'). We will say that $\mathscr{P}(a, b)$ holds if and only if

> a is the gödel number of a formula A(x), with x as its only free variable, and b is the gödel number of a proof of A(a).

Similarly, we will say that $\mathscr{R}(a, c)$ holds just in case

> a is the gödel number of a formula A(x), with x as its only free variable, and c is the gödel number of a proof of $\neg$A(a).

Remember that, in our notation, A(a) means that the numeral a has been substituted for x throughout A(x).

In the light of Chapter 9, it should not be difficult to believe that these predicates are R-computable. Therefore, we can use the main result from our previous chapter and let P(a, b) and R(a, c) be formulas which numeralwise express $\mathscr{P}(a, b)$ and $\mathscr{R}(a, c)$, respectively. We are now able to give a formula of N, call it G(g), which, because it is closed, is either true or false under the intended interpretation. However, we will show that neither $\vdash_N$ G(g) nor $\vdash_N \neg$G(g), if N is consistent.

Consider the formula G(a):

$$\forall b\,[\neg P(a, b) \vee \exists c(c \leqslant b\ \&\ R(a, c))]$$

EXERCISE 11.1.   Given the intended meanings of the symbols, (i) state in English what $\forall a$G(a) expresses: (ii) is $\forall a$G(a) true? Assume that N is consistent. ∎

This formula, G(a), has a gödel number, say g. Now, the formula we need is G(g):

$$\forall b\,[\neg P(g, b) \vee \exists c(c \leqslant b\ \&\ R(g, c))]$$

EXERCISE 11.2.   Given the intended meanings of its symbols, (i) state in English what G(g) means; (ii) is G(g) true? ∎

THEOREM 11.1. (Gödel-Rosser)   If N is consistent, then neither $\vdash_N$ G(g) nor $\vdash_N \neg$G(g).

Proof. To show that neither G(g) nor $\neg$G(g) is provable, we proceed by informal *reductio ad absurdum* in each case.

   (I) Assume (i) that N is consistent, and (ii) that $\vdash_N$ G(g). By (ii), we know that for some k, k is the gödel number of a proof of G(g), and $\mathscr{P}(g, k)$. Hence, by the expressibility of $\mathscr{P}(a, b)$, $\vdash_N$ P(g, k). But, by (i) and (ii), we know there is no proof of $\neg$G(g). Therefore, no number is the gödel number of a proof of $\neg$G(g), and it follows that $\mathscr{R}(g, 0), \mathscr{R}(g, 1)$ ..., and so on, are all false. By the expressibility of $\mathscr{R}(a, c)$, we get $\vdash_N \neg$R($g, 0$), $\vdash_N \neg$R($g, 1$) ...
   Since A(0), A(1),..., A(k) $\vdash \forall x(x \leqslant k \supset$ A(x)) (*166(a)),

$$\vdash_N \forall c(c \leqslant k \supset \neg R(g, c)).$$

Thus, with $\vdash_N$ P(g, k) and $\exists$-intro, we have

$$\vdash_N \exists b[P(g, b)\ \&\ \forall c(c \leqslant b \supset \neg R(g, c))].$$

Hence, by easy predicate logic, $\vdash_N \exists b[P(g, b)\ \&\ \neg \exists c(c \leqslant b\ \&\ R(g, c))]$ and thus

$$\vdash_N \neg \forall b[\neg P(g, b) \vee \exists c(c \leqslant b\ \&\ R(g, c))]$$

But this is $\vdash_N \neg$G(g), which yields a contradiction, given (i) and (ii). Therefore, if N is consistent, not $\vdash_N$ G(g).

   (II) Assume (i) that N is consistent and (iii) that $\vdash_N \neg$G(g). By (iii), for some

k, k is the gödel number of a proof of $\neg G(g)$. Hence, $\mathcal{R}(g, k)$ is true, so by the expressibility of $\mathcal{R}(a, b)$, $\vdash_N R(g, k)$. Applying $A(t) \vdash \forall x[x \geq t \supset \exists y(y \leq x \,\&\, A(y)]$ where t is a term not containing x free and t is free for x in $A(x)$, (*168), we can show that

$$\vdash_N \forall b[b \geq k \supset \exists c(c \leq b \,\&\, R(g, c))].$$

But, by (i) and part (I), there is no proof of $G(g)$, so that $\mathcal{P}(g, 0), \mathcal{P}(g, 1) \ldots$ are all false. Again, by expressibility, $\vdash_N \neg P(g, 0)$, $\vdash_N \neg P(g, 1), \ldots$ and so on. Applying $A(0), A(1), \ldots, A(k-1) \vdash \forall x(x < k \supset A(x))$ (*166) we see that

$$\vdash_N \forall b[b < \underline{k} \supset \neg P(g, b)].$$

Using the two previously displayed formulas and, where t is a term not containing x free, $\forall x[x < t \supset A(x)]$, $\forall x[x \geq t \supset B(x)] \vdash \forall x[A(x) \lor B(x)]$ (*169), we fine that

$$\vdash_N \forall b[\neg P(g, b) \lor \exists c(c \leq b \,\&\, R(g, c))].$$

But this is $\vdash_N G(g)$, which contradicts (i), given (iii). Hence, if N is consistent, not $\vdash_N \neg G(g)$. ∎

This concludes our proof of the incompleteness of the formal theory N. There is no doubt that it constitutes one of the most profound and fertile discoveries in mathematics. We shall demonstrate this in what follows by expounding some of its consequences and related results.

The unprovability of consistency

The first of the consequences we prove concerns the consistency of N, and is usually known as 'Gödel's Second Theorem'. In the proof of Theorem 11.1, we showed that

(A)   If {N is consistent} then [G(g) is unprovable].

It is not difficult to see that, via our gödel numbering and previous definitions, we can express both {...} and [...] by suitable formulas in N. For {...}, let c be the gödel number of the formula, with the one free variable a, $1 = 0 \,\&\, a = a$. Then, since $\vdash_N \neg 1 = 0$, we can express the consistency of N in N by $\forall b \neg P(c, b)$. This says that there is no proof in N of $1 = 0 \,\&\, c = c$. Call this formula, 'Consis.' [...] is obviously expressed in N by the formula $\forall b \neg P(g, b)$, which asserts that there is no proof in N of $G(g)$.

The proof of (A) presented above used only principles and methods whose formal counterparts belong to N. This is not obvious, but Hilbert and Bernays (1939) carried out the formalization of the proof of (A) in a system equivalent to N. Their formalization is far too lengthy for us to go through here. However, if the proof of (A) were carried out in N, we should have

$$\vdash_N \text{Consis} \supset \forall b \,\neg P(g, b).$$

Now, suppose we could prove the consistency of N in N, that is, $\vdash_N$ Consis. By *modus ponens*, this would yield $\vdash_N \forall b \,\neg P(g, b)$. But, by easy predicate logic, $\vdash_N \forall b \,\neg P(g, b) \supset G(g)$. With another use of *modus ponens*, we should have $\vdash_N G(g)$. By Theorem 11.2, this is impossible if N is consistent. Thus, we have proved the following:

THEOREM 11.2 (Gödel's Second Theorem).   If N is consistent, not $\vdash_N$ Consis. ∎

This establishes that it is impossible to prove the consistency of N within N. Hence, a consistency proof for N will have to use methods stronger than any formalizable in N. This result is commonly held to refute Hilbert's program. Whether it does so depends on what exactly counts as finitistic reasoning.

Developments using the diagonal function

In case the reader suspects that the incompleteness of the system N can be overcome by simply adding more axioms, including $G(g)$, for example, we conclude this chapter with some results which show why this is impossible. This will illuminate our earlier discoveries by setting them in a broader context. Where we can, we shall prove results for other formal theories besides N.

In what follows, let T be any first-order formal theory with the same symbols as N and for which the usual properties of identity (*100-*102 and the Replacement Theorem) are provable. Thus, in T we have the same underlying logic as in N. Also, the formal symbol '=' functions (in T) like ordinary identity. However, the formal symbols '+', '·', and '′' *need not be* the formal analogues of addition, multiplication and successor on the natural numbers.

First, we define the Diagonal function, $\text{Diag}(n)$ and the predicate, $\text{Prov}_T(n)$.

DEFINITION 11.1. $\text{Diag}(n)$ is the function such that, if n is the gödel number of a formula $A(x)$, with exactly the one free variable x, then the function $\text{Diag}(n)$ has as its value the gödel number of the formula $A(n)$; otherwise, $\text{Diag}(n) = n$. ∎

DEFINITION 11.2. $\text{Prov}_T(n)$ is true if n is the gödel number of a theorem of T, and otherwise, false. ∎

THEOREM 11.3. $\text{Diag}(n)$ is an R-computable function. ∎

EXERCISE 11.3.   Prove Theorem 11.3.

(Hint: assume that you have a machine which decides whether n is the gödel number of a formula with exactly the one free variable x (cf. Chapter 9), and proceed on that assumption.) ∎

THEOREM 11.4.   If T is consistent and Diag(n) is representable in T, then Prov$_T(n)$ is not expressible in T.

Proof. Suppose that Diag(n) is representable and Prov$_T(n)$ is expressible in T. Then there are formulas D(a, b) and P(b) such that,

(1)   if Diag(n) = m then $\vdash_T$ D(n, m), and

(2) $\vdash_T \exists! \text{b} \text{D}(n, \text{b})$;

and

(3)   if Prov$_T(n)$ is true then $\vdash_T$ P(n), while

(4)   if Prov$_T(n)$ is false then $\vdash_T \neg$P(n).

Consider the formula Q(a):

$$\forall \text{b}(\text{D}(\text{a}, \text{b}) \supset \neg\text{P}(\text{b})).$$

Let this formula have q as its gödel number. Now, the formula we need is Q(q):

$$\forall \text{b}(\text{D}(q, \text{b}) \supset \neg\text{P}(\text{b})).$$

EXERCISE 11.4.   Given the intended meanings of the symbols, express Q(a) and Q(q) in English, and say whether they are true. ■

If the gödel number of Q(q) is d, then Diag(q) = d and, by (1), $\vdash_T$D(q, d). Now, either $\vdash_T$ Q(q) or not $\vdash_T$ Q(q). If not $\vdash_T$ Q(q), then Prov$_T(d)$ is false, and (4) yields $\vdash_T \neg$P(d). However, if $\vdash_T$ Q(q), then $\vdash_T \forall$b(D(q, b) $\supset \neg$P(b)). Since $\vdash_T$ D(q, d), we have $\vdash_T \neg$P(d). Thus, in either case $\vdash_T \neg$P(d).

   Now, from $\vdash_T$ D(q, d) and (2), we have $\vdash_T$ D(q, b) $\supset$ b = d by easy predicate logic. But, since $\vdash_T \neg$P(d), $\vdash_T$ b = $d \supset \neg$P(b). Hence, $\vdash_T$ D(q, b) $\supset \neg$P(b). Rule 9 gives $\vdash_T \forall$b(D(q, b) $\supset \neg$P(b)), in other words, $\vdash_T$ Q(q). So, Prov$_T(d)$ is true, and by (3), $\vdash_T$ P(d). But we have already shown that $\vdash_T \neg$P(d). ■

   Now suppose we have an R-machine which decides whether Prov$_T(n)$ is true or false and prints 0 if it is true and 1 if it is false. Such a machine computes the *characteristic function*, C$_{\text{Prov}_T}(n)$, of the predicate Prov$_T(n)$ (see Definition 9.1). It is easy to show

THEOREM 11.5.   If the characteristic function C$_{\text{Prov}_T}(n)$ is representable in T by the formula B(a, b), then Prov$_T(n)$ is expressible by B(a, 0). ■

EXERCISE 11.5.   Prove Theorem 11.5. ■

   This yields:

THEOREM 11.6.   If T is consistent and every total R-computable function is representable in T, then $Prov_T(n)$ is not expressible in T. Hence, $C_{Prov_T}(n)$ is not R-computable. ∎

EXERCISE 11.6.   Prove Theorem 11.6. ∎

R-decidability

Clearly, if, for a theory T satisfying our conditions, there is no R-machine which will print 0 when and only when n is the gödel number of a theorem of T, there is no machine which can decide whether an arbitrary formula is provable in T or not. We will call such a theory 'R-undecidable'.

DEFINITION 11.3. T is *R-decidable* if $C_{Prov_T}(n)$ is R-computable and *R-undecidable* otherwise. Furthermore, T is *essentially R-undecidable* if and only if T and every consistent extension of T is R-undecidable. ∎

Recall that T' is an *extension* of a formal theory T if T' has the same symbols as T and every theorem of T is a theorem of T'.
    A number of interesting results now follow immediately.

THEOREM 11.7.   If N is consistent, then N is R-undecidable. ∎

EXERCISE 11.7.   Prove Theorem 11.7. ∎

THEOREM 11.8.   No consistent extension of N is R-decidable. ∎

EXERCISE 11.8.   Prove Theorem 11.8. ∎

THEOREM 11.9. N is essentially R-undecidable. ∎

EXERCISE 11.9.   Prove Theorem 11.9. ∎

Tarski's Theorem

Historically speaking, it sometimes happens that a mathematical notion which was introduced for one purpose is later redefined in order to serve some other purpose. This has certainly happened with the notion of *formal theory*. When we introduced the notion in Chapter 4, there were two elements motivating the definition: finitism and formalism. It was the first of these, finitism, which imposed the requirement on our definition that 'It must be decidable which formulas are axioms'. If we remove this requirement for the remainder of this chapter, keeping everything else as it was in our original definition, we are able to present a very important result of Tarski's.

DEFINITION 11.4.   We call a theory **T** *axiomatizable* if there is some R-decidable subset of formulas of **T** whose consequences (in **T**) are exactly the theorems of **T**. ∎

N is, of course, axiomatizable, but we shall see below that the theory **A** is not.

THEOREM 11.10.   If **T** is axiomatizable and complete, **T** is R-decidable.

Proof. If **T** is axiomatizable, we can enumerate the proofs and hence the theorems of **T**. If **T** is complete, then, for any closed formula A, either $\vdash_\mathbf{T} A$ or $\vdash_\mathbf{T} \neg A$.

EXERCISE 11.10.   Complete the proof of Theorem 11.10. ∎

A very powerful theorem is now available.

THEOREM 11.11.   There is no consistent, complete, axiomatizable extension of **N**. ∎

EXERCISE 11.11.   Prove Theorem 11.11. ∎

We might express this by saying that **N** is 'essentially incomplete'.

Now, let **A** (for 'Arithmetic') be the extension of **N** having as its axioms exactly those formulas which are true in the intended interpretation. Notice that **A** has the same symbols and formulas as **N**. We assume **A** is consistent. Clearly, if $\vdash_\mathbf{N} K$ then $\vdash_\mathbf{A} K$, so all total R-computable functions are representable in **A** and Theorem 11.4 applies to **A**. The reader should check the proof to be sure. It now follows that

THEOREM 11.12.   **A** is not R-decidable. ∎

EXERCISE 11.12.   Prove Theorem 11.12. ∎

THEOREM 11.13.   **A** is not axiomatizable. ∎

EXERCISE 11.13.   Prove Theorem 11.13. ∎

THEOREM 11.14. (Tarski 1936)   Let Tr(n) hold if n is the gödel number of a formula of **N** (or of **A**), which is true on the intended interpretation. Tr(n) is not expressible in **A**. ∎

EXERCISE 11.14.   Prove Theorem 11.14.

(Hint: $\text{Prov}_\mathbf{A}(n)$ if and only if $\text{Tr}(n)$.) ∎

We may express Tarski's result by saying that, in the language of arithmetic, there is no formula F(x), with one free variable, x, which is true of just those numbers which are gödel numbers of truths of arithmetic, or more briefly still, that the notion of arithmetical truth cannot be expressed or defined in arithmetic itself.

12

UNDECIDABLE PROBLEMS

Having shown that axiomatizable extensions of **N** remain incomplete and undecidable — we might put this roughly by saying that we cannot escape Gödel's results by strengthening **N** — it is natural to ask if related results can be obtained for weaker theories, subtheories, of **N**. We now present a subtheory of **N** for which many of the preceding results hold and which has some other interesting properties, too. This is the formal theory Q^+.

The formal theory Q^+

Q^+ has the axioms 1–12, as does the formal theory **N**. Therefore, it has the same underlying logic. Q^+ also has the following 14 axioms:

13. $a = a$

14. $a = b \supset b = a$

15. $a = b \supset (b = c \supset a = c)$

16. $a = b \supset a' = b'$

17. $a = b \supset (a + c = b + c \, \& \, c + a = c + b)$

18. $a = b \supset (a \cdot c = b \cdot c \, \& \, c \cdot a = c \cdot b)$

19. $a' = b' \supset a = b$

20. $\neg \, 0 = a'$

21. $\neg a = 0 \supset \exists b(a = b')$

22. $a + 0 = a$

23. $a + b' = (a + b)'$

24. $a \cdot 0 = 0$

25. $a \cdot b' = a \cdot b + a$

26. $b \neq 0 \supset \exists! q \exists! r(a = b \cdot q + r \, \& \, r < b)$.

Axioms 13–25 are due to Robinson (1950). Axiom 26 is added to simplify the proof of Theorem 12.4, below.

$\mathbf{Q}^+$ is clearly a sub-theory of $\mathbf{N}$; all the axioms of $\mathbf{Q}^+$ are theorems of $\mathbf{N}$. (It is an exercise if you doubt it.) It is also a formal theory in which the usual properties of identity are provable; axioms 13–18 are sufficient to guarantee this. Again, if you don't believe it, take it as an exercise.

The representability of R-computable functions in $\mathbf{Q}^+$

It requires only small modifications to our proof of Theorem 10.2 to show that all total R-computable functions are representable in $\mathbf{Q}^+$. In particular we need the following.

LEMMA 12.1.

(i) The relations $a = b$ and $a < b$ are numeralwise expressed in $\mathbf{Q}^+$ by the formulas a = b and $\exists c(c' + a = b)$, respectively.

(ii) The functions $a', a + b$, and $a \cdot b$ are numeralwise represented in $\mathbf{Q}^+$ by the formulas a′ = b, a + b = c and a·b = c, respectively.

Proof. Exactly as for $\mathbf{N}$, with appropriate changes of axiom numbers. See answers to Exercises 10.1 and 10.4. ∎

LEMMA 12.2.   Gödel's β-function is strongly representable in $\mathbf{Q}^+$.

Proof. Clearly, $\vdash_{\mathbf{Q}^+} \neg 0 = (i' \cdot d)'$ by axiom 20. Hence, by axiom 26,

$$\vdash_{\mathbf{Q}^+} \exists! w \exists u (c = (i' \cdot d)' \cdot u + w \ \& \ w < (i' \cdot d)').$$

The remainder is exactly as for Theorem 10.1. ∎

We also need, at steps 5 and 6, respectively, in Exercise 10.13, the following:

LEMMA 12.3.

(i) $\vdash_{\mathbf{Q}^+} a \leqslant n \supset a = 0 \lor a = 1 \lor \ldots \lor a = n$

(ii) $\vdash_{\mathbf{Q}^+} a \leqslant n \lor n \leqslant a.$ ∎

EXERCISE 12.1.   Prove Lemma 12.3 (i).

(Hint: this is a long and difficult exercise which the reader may skip if he wishes. Otherwise, the proof is by induction in the metalanguage.) ∎

EXERCISE 12.2.   Prove, for any numeral n, that:

(i) $\vdash_{Q^+} a \leqslant n \supset a \leqslant n'$;

(ii) $\vdash_{Q^+} a' + n = a + n'$; and

(iii) $\vdash_{Q^+} n < a \supset n' \leqslant a$.

Then use (i) through (iii) to prove Lemma 12.3 (ii) by induction in the meta-language.

(Hint: use Lemma 12.3 (i) for (i): (iii) follows from (ii). Again, this is a long exercise which the reader may skip if he wishes.) ∎

These results are sufficient to establish:

THEOREM 12.4.   All total R-computable functions are numeralwise represent-able in Q^+. ∎

It is left to the reader to check the details. Theorem 12.4 quickly yields the fol-lowing results.

THEOREM 12.5.   (i) Diag(n) is representable in Q^+. Hence, if Q^+ is consistent, (ii) Prov$_{Q^+}(n)$ is not expressible in Q^+; (iii) Q^+ is R-undecidable; and (iv) Q^+ is essentially R-undecidable. ∎

EXERCISE 12.3.   Prove Theorem 12.5 (i)–(iv). ∎

Some results of Tarski's

What is especially interesting about Q^+ is that it is *finitely axiomatizable*, i.e. all its theorems are derivable from a finite number of proper axioms (13–26 above). This fact yields some important results due to Tarski (cf. Tarski–Mostowski–Robinson 1953).

EXERCISE 12.4.

(i)   How many proper *axioms* has the formal theory N?

(ii)   Does your answer tell you whether N is finitely axiomati*zable*? ∎

First, some definitions for first-order theories, T_1, T_2:

DEFINITION 12.1. T_2 is called a *finite extension* of T_1 if, for some *finite* set of formulas E, every theorem of T_2 is derivable from a set of axioms which are either theorems of T_1 or are in E. ∎

DEFINITION 12.2.   **T** is called the *union* of **T**$_1$ and **T**$_2$ provided that: (i) the set of symbols of **T** is the union of the sets of symbols of **T**$_1$ and **T**$_2$, in other words, a symbol belongs to **T** iff it belongs either to **T**$_1$ or to **T**$_2$; and (ii) any theorem of **T** is derivable from a set of formulas, each of which is provable either in **T**$_1$ or in **T**$_2$. ▪

DEFINITION 12.3.   **T**$_1$ and **T**$_2$ are called *compatible* if their union is consistent. ▪

EXERCISE 12.5.   Let **D** be Presburger's formal theory.

   (i)   Is **N** a finite extension of **D**?

   (ii) What is the union of **D** and **N**?

   (iii) Let **N**$^+$ be **N** plus the additional axiom $G(g)$ and let **N**$^\neg$ be **N** plus the additional axiom $\neg G(g)$. Are **N** and **N**$^+$ compatible? Are **N**$^+$ and **N**$^\neg$ compatible? ▪

Now we prove some of Tarski's results.

THEOREM 12.6.   Let **T**$_1$, **T**$_2$ be first-order theories with the same symbols, such that **T**$_2$ is a finite extension of **T**$_1$. Then, if **T**$_2$ is R-undecidable, so is **T**$_1$. ▪

EXERCISE 12.6.   Prove Theorem 12.6.

(Hint: use Definition 12.1 and the Deduction Theorem.) ▪

THEOREM 12.7.   Let **T** be a first-order theory with the same symbols as **Q**$^+$ such that **T** is compatible with **Q**$^+$. Then **T** is R-undecidable. ▪

EXERCISE 12.7.   Prove Theorem 12.7.

(Hint: use Definition 12.3, Theorem 12.5 (iv) and Theorem 12.6.) ▪

   We now let **Pr** be the predicate calculus having the same symbols as **Q**$^+$ (or **N**). **Pr** is the formal theory having *no* proper axioms, but only axioms of the forms 1~12.

THEOREM 12.8.   **Pr** is R-undecidable. ▪

EXERCISE 12.8.   Prove Theorem 12.8. ▪

Church's Theorem

A formula is called *logically valid* if it is true whatever interpretation is given to its predicate letters, function letters, and constants. Although we have not done so, it is possible to prove that any formula expressed in the symbolism of Q^+ is logically valid if and only if it is provable in **Pr** (cf. Kleene 1967, Ch. VI). Assuming this, our previous theorem immediately yields *Church's Theorem* (1936b).

THEOREM 12.9 (*Church's Theorem*).   There is no R-machine which can decide whether an arbitrary formula is logically valid. ∎

Accepting Church's Thesis, that all intuitively computable functions are R-computable, Church's Theorem means that, in general, to tell whether a formula is logically valid will require ingenuity and inventiveness; no machine can tell you.

EXERCISE 12.9.   Prove Theorem 12.9. ∎

The halting problem

We conclude with a direct proof of a result which is closely related to our previous results and which uses methods which have a clear affinity to some we have used previously. We ask this question: 'For an arbitrary R-machine M_i, can we decide whether M_i stops when applied to a given argument a?' We shall shortly see that there is no decision procedure for this problem. This is usually expressed by saying that the *halting problem* is unsolvable. To show the unsolvability of the halting problem for R-machines, we first introduce a gödel numbering for programs. We then describe a *universal* R-machine and finally we give the proof.

Gödel numbering programs

We code the instructions of a program $I_0, \dots, I_h$, where $I_h = \text{STOP}$, thus:

$$\text{code}(Z_m) = 3m - 1$$

$$\text{code}(S_m) = 3m + 1$$

$$\text{code}(J_{i,j,k}) = 3c \text{ where } c < (2h + 1) \cdot (2h + 2) \cdot (2h + 3),$$

$$\text{rm}(c, 2h + 1) = i,$$

$$\text{rm}(c, 2h + 2) = j, \text{ and}$$

$$\text{rm}(c, 2h + 3) = k; \text{ and}$$

$$\text{code}(\text{STOP}) = 1.$$

Let p_i be the ith prime and $p_0 = 2$. Then the gödel number of a program,

$I_0, \ldots, I_h$, is simply

$$p_0^{\text{code}(I_0)} \cdot p_1^{\text{code}(I_1)} \cdot \ldots \cdot p_h^{\text{code}(I_h)}$$

EXERCISE 12.10.   Show that the c used in coding $J_{i,j,k}$ is *unique*. ■

It is obviously a 'mechanical' matter to retrieve a program from its gödel number and it should also be clear that there is an R-machine which can decide whether, under this coding, i is the gödel number of a program. (See the exercises to Chapter 9 and the R-machine for deciding whether v is the gödel number of a term, pp. 104 f. if it is not clear.)

A universal R-machine

Let $\Phi(i, a)$ be the partial function which takes the value computed by M_i, the R-machine with gödel number i, when M_i is applied to argument a. Hence, $\Phi(i, a)$ is defined for just those pairs (i, a) such that M_i will compute a value when applied to argument a.

We shortly give a flow diagram for a program which computes $\Phi(i, a)$ if it is defined and does not stop otherwise. First, there are some preliminaries.

(i)   Let R_1 contain i and R_2 contain a.

(ii)   The first decision is 'does R_1 contain the gödel number of an R-machine?' This is so provided i is of the form

$$p_0^{x_0} \cdot p_1^{x_1} \cdot \ldots \cdot p_h^{x_h}$$

where $x_h = 1$; $x_j > 1$ for $0 \leqslant j < h$; and if $3 | x_j$ then $\text{rm}(x_j/3, 2h + 3) \leqslant h$. This can obviously be tested by earlier programs, so we give no further details. Take it as an exercise, if you doubt it.

(iii)   If R_1 contains the gödel number of a machine of the form

$$p_0^{x_0} \cdot p_1^{x_1} \cdot \ldots \cdot p_h^{x_h},$$

then x_j is the code number of the jth instruction of the machine, M_i.

(a)   Let R_3 contain successively the numbers (*not* the code numbers) of the instructions that M_i would carry out when applied to argument a.

(b)   Where R_3 contains the number d for the tth time, let R_4 contain a number of the form

$$p_1^{y_1} \cdot p_2^{y_2} \cdot \ldots \cdot p_k^{y_k}.$$

Here, y_j is the number that M_i, when applied to argument a, would have in its jth register when about to execute instruction d for the tth time. So R_4 starts by containing 3^a.

(iv) We shall write

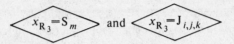

as shorthand for 'Is the exponent of p_{R_3} of R_1 the code number for Z_m?' It is easy to see that this can be tested by earlier programmes.

$$\langle x_{R_3}=S_m \rangle \quad \text{and} \quad \langle x_{R_3}=J_{i,j,k} \rangle$$

are interpreted similarly.

Now we can display $R_1 := \Phi(R_1, R_2)$, a flow diagram for $\Phi(i, a)$.

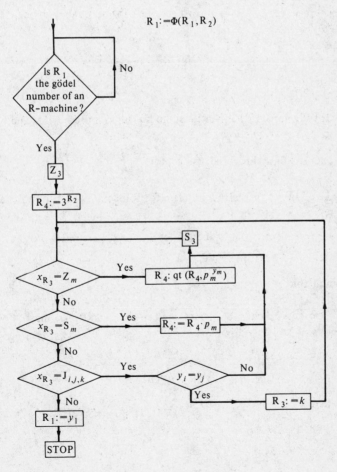

EXERCISE 12.11.　　Use the program which copies the contents of R_1 into R_3 (see p. 91) to check how this program works. If R_2 contains the number 2, what are the successive different contents of R_3 and R_4?

(Hint: they start out containing 0 and 3^2, respectively.) ∎

The unsolvability of the halting problem

To show that the halting problem for R-machines is unsolvable we shall assume Church's Thesis, that all intuitively computable functions are R-computable. A lemma using a kind of diagonalization will then quickly yield the required result.

LEMMA 12.10.　　The function $f(a)$ defined by

$$f(a) = \begin{cases} \Phi(a, a) + 1 & \text{if } \Phi(a, a) \text{ is defined} \\ 0 & \text{otherwise,} \end{cases}$$

is *not* computable. (Notice that $f(a)$ is a total function.) ∎

EXERCISE 12.12.

(i)　Prove Lemma 12.10 (Hint: assume $f(a)$ is computed by M_p and hence derive a contradiction.)

(ii)　Is $f(a)$ computable for all a except p?

THEOREM 12.11.　　The halting problem for R-machines is unsolvable. Specifically, the function $g(a)$ defined by

$$g(a) = \begin{cases} 0 & \text{if } M_a \text{ computes a value for argument } a \\ 1 & \text{otherwise} \end{cases}$$

is *not* computable.

Proof. Suppose it is computable. Then we can devise an R-machine to compute $f(a)$ of Lemma 12.10 as follows:

(i)　compute $g(a)$;

(ii)　if $g(a) = 1$, then $f(a) = 0$, then STOP;

　　if $g(a) = 0$, go to (iii);

(iii)　compute $\Phi(a, a)$;

(iv)　$f(a) = \Phi(a, a) + 1$,

(iii) is possible by the existence of a Universal R-machine. Since, by Lemma 12.10, f(a) is *not* computable, neither is g(a). Therefore, the halting problem is unsolvable. ■

EXERCISE 12.13.   Show that if **N** is R-undecidable, the halting problem is unsolvable. ■

APPENDIX:
ANSWERS TO EXERCISES

Chapter 1

1.1.

(i) $1 \cdot a = a$, so there is an integer c such that
$1 \cdot c = a$, so by the definition $1 | a$.

(ii) $a \cdot 1 = a$, so there is an integer c such that
$a \cdot c = a$, so by the definition $a | a$.

(iii) $a \cdot 0 = 0$, so there is an integer c such that
$a \cdot c = 0$, so by the definition $a | 0$.

(iv) Assume that $a | b$ and $b | c$. Then there are integers n and m such that $a \cdot n = b$ and $b \cdot m = c$. Substituting for b we get $(a \cdot n) \cdot m = c$, hence $a \cdot (n \cdot m) = c$, so $a | c$.

(v) Assume that $a | b$. Then, for some n, $a \cdot n = b$. Clearly $(a \cdot n) \cdot c = b \cdot c$, so $(a \cdot c) \cdot n = b \cdot c$, so $a \cdot c | b \cdot c$.

(vi) Assume that $a | b$ and $a | c$. Suppose $a \cdot x = b$ and $a \cdot y = c$. Then $m \cdot b + n \cdot c = m \cdot (a \cdot y) = a \cdot (m \cdot x + n \cdot y)$, so $a | m \cdot b + n \cdot c$.

(vii) Assume $a > 1$ and that there are integers n and m such that $a \cdot n = b$ and $a \cdot m = b + 1$. Since $b < b + 1$, $a \cdot n < a \cdot m$ and therefore $n < m$ (as $a \neq 0$). But then there is an $x \geqslant 1$ such that $n + x = m$, so $b + 1 = a \cdot m = a \cdot n + a \cdot x = b + a \cdot x$. Hence, $1 = a \cdot x$ which contradicts the assumption that $a > 1$.

1.2. $1404 = 2 \cdot 2 \cdot 3 \cdot 3 \cdot 3 \cdot 13$, and
$1560 = 2 \cdot 2 \cdot 2 \cdot 3 \cdot 5 \cdot 13$;
so GCD $= 2 \cdot 2 \cdot 3 \cdot 13 = 156$.

1.3.   If $p|b$ the theorem is true. If $p\!\not|b$ then $(p, b) = 1$ since the only (positive) divisors of p are p and 1. Hence, by Theorem 1.4 there are integers x, y such that $1 = bx + py$. Multiplying by a we have $a = abx + apy$. But, by hypothesis $p|ab$ and clearly $p|p$, so $p|a$. (By Theorem 1.1 (vi), etc.)

1.4.

   (i)   If p is prime and $p|a(bc...n)$ then, by Theorem 1.5, $p|a$ or $p|(bc...n)$. If $p|b(c...n)$ then $p|b$ or $p|(c...n)$, etc.

   (ii)   Suppose $p|k$ and k is prime. Then $p = k$.

1.5.   Assume there are only finitely many primes. Let $2, 3, 5, 7,..., p_n$ be the complete list of primes up to and including the last prime p_n, and let

$$N = (2 \cdot 3 \cdot 5 \cdot 7 \cdot ... \cdot p_n) + 1.$$

Clearly this number is not divisible by 2. (If $2|N$, then since $2|(2 \cdot 3 \cdot 5 \cdot ... \cdot p_n)$, we find $2|(N - (2 \cdot 3 \cdot 5 \cdot ... \cdot p_n))$ by Theorem 1.9 (ii); hence, $2|1$ which is impossible.) Similarly $3\!\not|N, 5\!\not|N,..., p_n\!\not|N$. Now, either N is prime or it is composite: if it is prime there is a prime $> p_n$; if it is composite it is divisible by *some* prime and, again, there is a prime $> p_n$. So, for any prime p_n one can always find another, p_m, in the interval $p_n < p_m \leqslant N$. This shows there is no largest prime, so there must be infinitely many distinct primes.

1.6.

   (i) $a = n \cdot c + r$ and $b = m \cdot c + r$, so $a - b = c \cdot (n - m)$.

   (ii)   An immediate corollary of (i).

1.7.

   (i)   By Theorem 1.4 we have for some m and n (since p and q are relatively prime) $1 = pm + qn$. So $a = pma + qna$. Let $a = pb$ and $a = qc$, then $a = pmqc + qnpb = pq(mc + nb)$.

   (ii)   An immediate corollary of (i).

1.8.

   (i)   No; the ith members are not equal.

   (ii)   No; the sequences are of different length.

   (iii)   No.

1.9.

(i) As c takes successively the values $0,\ldots, 12$, $rm(c, 2)$ takes successively the values $0, 1, 0, 1,\ldots$ and $rm(c, 6)$ the values $0, 1, 2, 3, 4, 5, 0, 1, 2, 3, 4, 5$. So the six couples $\langle rm(c, 2), rm(c, 6)\rangle$ are

$$\langle 0, 0\rangle, \langle 1, 1\rangle, \langle 0, 2\rangle, \langle 1, 3\rangle, \langle 0, 4\rangle, \langle 1, 5\rangle.$$

(ii) $rm(c, 3)$ takes successively the values $0, 1, 2, 0, 1, 2,\ldots$ and $rm(c, 4)$ the values $0, 1, 2, 3, 0, 1, 2, 3,\ldots$. This produces *twelve* couples $\langle rm(c, 3), rm(c, 4)\rangle$.

Chapter 2

2.1.

(ii)   1. $a \cdot 1 = a$.   true for all a.

         2. $a \cdot 1 = a \supset \exists c(ac = a)$ logical truth, $P(1) \supset \exists c P(c)$.

         3. $\exists c(ac = a)$ 1, 2, and *modus ponens*.

         4. $a|a \sim \exists c(ac = a)$ definition.

         5. $(a|a \sim \exists c(ac = a)) \supset (\exists c(ac = a) \supset a|a)$ truth about propositions.

         6. $\exists c(ac = a) \supset a|a$ 4, 5, and *modus ponens*.

         7. $a|a$ 3, 6, and *modus ponens*.

(iii)   1. $a \cdot 0 = 0$ true for all a.

         2. $a \cdot 0 = 0 \supset \exists c(ac = 0)$ logical truth, $P(0) \supset \exists c P(c)$.

         3. $\exists c(ac = 0)$ 1, 2, and *modus ponens*.

         4. $a|0 \sim \exists c(ac = 0)$ definition.

         5. $(a|0 \sim \exists c(ac = 0)) \supset \exists c(ac = 0) \supset a|0$ truth about propositions.

         6. $\exists c(ac = 0) \supset a|0$ 4, 5, and *modus ponens*.

         7. $a|0$ 3, 6, and *modus ponens*.

2.2.   Instead of $(A \sim B) \supset (B \supset A)$, this uses $(A \sim B) \supset (A \supset B)$.

2.3.

(v)   1. $a|b$ assumption.

         2. $a|b \sim \exists n(an = b)$ definition.

3. $\exists n(an = b)$ 2, 1, and *modus ponens*.

4. $an = b$ assumption.

5. $(an)c = bc$ by $r = s \supset rt = st$, 4 and *modus ponens*.

6. $(an)c = (ac)n$ by $(an)c = a(nc)$ (Associativity), $nc = cn$ (Commutativity), $r = s \supset tr = ts$, *modus ponens*, and $r = s \supset (s = t \supset r = t)$ (Transitivity of =).

7. $(ac)n = bc$ by 6, 5, and $r = s \supset (r = t \supset s = t)$.

8. $\exists d((ac)d = bc)$ from 7 and $P(n) \supset \exists dP(d)$.

9. $ac \mid bc$ by 8, definition.

(vi) 1. $a \mid b$ & $a \mid c$ assumption.

2. $ap = b$ assumption.

3. $aq = c$ assumption.

4. $map = mb$ 2, 3, and $r = s \supset tr = ts$ with *modus ponens*.

5. $naq = nc$ 2, 3, and $r = s \supset tr = ts$ with *modus ponens*.

6. $map + naq = mb + nc$ by $r = s \supset r + t = s + t$, $r = s \supset t + r = t + s$, and Transitivity of identity.

7. $map + naq = amp + anq$ commutativity, etc.

8. $amp + anq = a(mp + nq)$ by $t(r + s) = tr + ts$.

9. $mb + nc = a(mp + nq)$ by Transitivity of identity, etc.

10. $a \mid mb + nc$ by 9, definition, etc.

2.4. 1. $A(0)$ & $\forall x(A(x) \supset A(x'))$ assumption.

2. $A(0)$ 1, by $(A \& B) \supset A$.

3. $\forall x(A(x) \supset A(x'))$ 1, by $(A \& B) \supset B$.

4. $x = 0 \lor x > 0$ truth about natural numbers.

5. $\forall y(y < 0 \supset A(y)) \supset A(0)$ 2, by $B \supset (A \supset B)$.

6. $x = 0 \supset [\forall y(y < x \supset A(y)) \supset A(x)]$.   5.

7. $x > 0 \supset \exists z(x = z')$ true.

8. $x > 0$ assumption.

9. $\exists z(x = z')$ 7, 8, *modus ponens.*

10. $x = z'$ assumption.

11. $\forall y(y < z' \supset A(y))$ assumption.

12. $z < z' \supset A(z)$ 11, $\forall$-elimination.

13. $z < z'$ true.

14. $A(z)$ 12, 13, *modus ponens.*

15. $A(z) \supset A(z')$ 3, $\forall$-elimination.

16. $A(z')$ 14, 15, *modus ponens.*

17. $\forall y(y < z' \supset A(y)) \supset A(z')$ 11, 16.

18. $x = z' \supset [\forall y(y < x \supset A(y)) \supset A(x)]$ 10, 17.

19. $\exists z(x = z') \supset [\forall y(y < x \supset A(y)) \supset A(x)]$ 9, 18.

20. $x > 0 \supset [\forall y(y < x \supset A(y)) \supset A(x)]$ 8, 19, and $[(A \supset B) \&$ $(B \supset C)] \supset (A \supset C)$.

21. $\forall y(y < x \supset A(y)) \supset A(x)$ 4, 6, 20, and Proof by cases.

22. $\forall x[\forall y(y < x \supset A(y)) \supset A(x)]$ 21.

23. $\forall x[\forall y(y < x \supset A(y)) \supset A(x)] \supset \forall x Ax$ Complete induction.

24. $\forall x Ax$ 22, 23, *modus ponens.*

25. $[A(0) \& \forall x(A(x) \supset A(x'))] \supset \forall x Ax$ 1, 24.

2.5.   If $x = 2$ there is no $y < x$ (since, by hypothesis $y > 1$) so $\forall y(y < x \supset F(y))$ is trivially true. If $x > 2$ there are $y < x$ (and $y > 1$).

2.6.   We wish to prove that if p is prime and $p \mid ab$ then $p \mid a$ or $p \mid b$. We start from $p \mid b \lor \neg p \mid b$ (Excluded middle) and do a proof by cases.
   Case 1   If $p \mid b$ then $p \mid a \lor p \mid b$, by $A \supset (B \lor A)$.
   Case 2   Assume $\neg p \mid b$. Then, since p is prime, $(p, b) = 1$ by the definition. Hence, by Theorem 1.4 there are integers x, y such that $1 = bx + py$. Assume $1 = bx + py$, then $a = abx + apy$. This follows by $r = s \supset tr = ts$, $a(r + s) = ar + as$, and the transitivity of $=$. Clearly $p \mid apy$ (Theorem 1.1 (ii), (v)). By assumption $p \mid ab$ so $p \mid a$ (by Theorem 1.1 (v), (v), and properties of $=$). Hence $p \mid a \lor p \mid b$ (by $B \supset (B \lor A)$).
   Note: if we wanted to prove this result without using negative integers, we

could replace Theorem 1.4 with the following: for natural numbers a, b, d with $(a, b) = d$ there are natural numbers x, y such that

$$d + bx = ay \lor d + ay = bx.$$

Assuming $\neg p \mid b$ a new proof would again show that $p \mid a \lor p \mid b$.

Chapter 3

3.1. Let $R(a, b)$ stand for $a < b$ where a and b are natural numbers. Then $\forall a \exists b R(a, b)$ reads 'for every number a there is a greater number b, which is *true*. But $\exists b \forall a R(a, b)$ reads, 'there is a number b such that for every number a, $a < b$' or 'there is a *greatest* number', which is *false*.

3.2.

(i) (P ⊃ (Q & ¬ P)) ⊃ (P ⊃ Q)

T	F	T	F	F	T	T	T	T	T
T	F	F	F	F	T	T	T	F	F
F	T	T	T	T	F	T	F	T	T
F	T	F	F	T	F	T	F	T	F

Tautology

(ii) (P ⊃ (Q & ¬ P)) ⊃ (Q ⊃ P)

F		T	T	T	T
F		T	F	T	T
T		F	T	F	F
T		T	F	T	F

Not a tautology

3.3.

(i) (P ⊃ Q) ⊃ ((Q ⊃ R) ⊃ (P ⊃ R))

T_4 T_2 $\underline{T_6}$ F_1 $\underline{F_7}$ T_3 F_5 F_2 T_4 F_3 F_4

(ii) (P ⊃ Q)⊃(¬ Q ⊃ ¬ P);   (¬ Q ⊃ ¬ P)⊃(P ⊃ Q)

$T_5 \, T_2 \, \underline{T_6} \, F_1 \, T_3 \, \underline{F_7} \, F_2 \, F_3 \, T_4$ $T_5 \, F_4 \, T_2 \, T_6 \, \underline{F_7} \, F_1 \, \underline{T_8} \, F_2 \, F_3$

(iii) (P ⊃ Q) ⊃ ((¬ P ⊃ Q) ⊃ Q)

$\underline{F_5}$ T_2 F_4 F_1 F_6 $\underline{T_7}$ T_3 F_4 F_2 F_3

(iv) (P ⊃ Q) ⊃ ((P ⊃ ¬ Q) ⊃ ¬ P)

    T_5 T_2 $\underline{T_6}$ F_1 T_5 T_3 T_6 $\underline{F_8}$ F_2 F_3 T_4

(v) (P ⊃ R) ⊃ ((Q ⊃ R) ⊃ ((P ∨ Q) ⊃ R))

    F_6 T_2 F_5 F_1 $\underline{F_9}$ T_3 F_5 F_2 F_7 T_4 $\underline{T_8}$ F_3 F_4

(vi) (¬ P ⊃ P) ⊃ P

    F_4 $\underline{T_5}$ T_2 F_3 F_1 $\underline{F_2}$

(vii) (P & (P ⊃ Q)) ⊃ Q

    T_3 T_2 T_4 T_3 $\underline{T_5}$ F_1 $\underline{F_2}$

(viii) P ⊃ (Q ⊃ (P & Q))

    T_2 F_1 $\underline{T_3}$ F_2 T_4 F_3 $\underline{F_5}$

(ix) (P ⊃ Q) ⊃ (Q ⊃ P)

    F_4 T_2 T_5 F_1 T_3 F_2 F_3

This formula is false when P is false and Q is true.

(x) ((P ⊃ Q) ⊃ (Q ⊃ R)) ⊃ (P ⊃ R)

    T_4 F_6 F_5 T_2 F_5 T_6 F_4 F_1 T_3 F_2 F_3

This is not a tautology (cf. 3.3 (i)). Brackets are important!

(xi) ¬ (P ⊃ Q) ⊃ (¬ P ⊃ Q)

    T_2 $\underline{T_4}$ F_3 F_4 F_1 T_6 $\underline{F_7}$ F_2 F_5

(xii)

((P ⊃ Q) ⊃ (R ⊃ S)) ⊃ ((¬ S ⊃ ¬ Q) ⊃ (P ⊃ ¬ R))

  T_6 T_{14} T_8 T_2 T_7 T_{13} T_{12} F_1 F_{10} T_{11} T_3 F_9 T_8 F_2 T_4 F_3 F_4 T_5

This is not a tautology.

All except (ix), (xi), and (xii) are tautologies and are sound. In (xii) step 8 was a guess. In such a case start the short truth-table method assigning *necessary* values as far as possible and when there is a choice make a guess. This will often yield a decision quite quickly. The alternative is to draw out the full truth-table of 16 lines!

3.5.

(i) (¬ ¬ P) ~ P

T	F	T	T	T
F	T	F	T	F

(ii) (¬ P ∨ Q) ~ (P ⊃ Q)

F	T	T	T	T		T
F	T	F	F	T		F
T	F	T	T	T		T
T	F	T	F	T		T

(iv) (P & Q) ~ ¬ (¬ P ∨ ¬ Q)

T	T	T	T	T	F		F	F
T	F	F	T	F	F		T	T
F	F	T	T	F	T		T	F
F	F	F	T	F	T		T	T

(vi) (P ⊃ Q) ~ ¬ (P & ¬ Q)

T	T	T	T	T		F	F
T	F	F	T	F		T	T
F	T	T	T	T		F	F
F	T	F	T	T		F	T

(vii) (P ~ Q) ~ ((P ⊃ Q) & (Q ⊃ P))

T	T	T	T	T	T	T	T	T	T	T
T	F	F	T	T	F	F	F	F	T	T
F	F	T	T	F	T	T	F	T	F	F
F	T	F	T	F	T	F	T	T	F	F

3.6. Eliminate each occurrence of ~ by using (vii), then eliminate each occurrence of & by using (IV), then eliminate each occurrence of ∨ by using (iii). Strings of two or more occurrences of ¬ may be collapsed using (i). The resulting formula has the same truth table as the original. (Why?)

3.7. For domain $D = \{1, 2\}$, the formula expands as follows:

(Step 1) $\forall a[R(a, 1) \vee R(a, 2)] \supset \exists b[R(1, b) \,\&\, R(2, b)]$

(Step 2) $\{[R(1, 1) \vee R(1, 2)] \,\&\, [R(2, 1) \vee R(2, 1)]\}$

$\supset \{[R(1, 1) \,\&\, R(2, 1)] \vee [R(1, 2) \,\&\, R(2, 2)]\}.$

The problem then simplifies to:

$$\{[P \quad \vee \quad Q] \,\&\, [R \quad \vee \quad S]\} \supset \{[P \,\&\, R] \vee [Q \,\&\, S]$$

$$\begin{array}{cccccccccccc} & T_4 & & T_2 & & T_4 & & F_1 & T_5 \; F_3 & F_5 & F_2 \; T_5 \; F_3 & F_5 \end{array}$$

By the 'guess' at step 5 we have found an assignment of truth values which makes the original formula false in D.

Chapter 4

4.1.

(i)   Yes. Let M be the set of human males and S the set of sons.
      $\forall a(a \in M \sim a \in S)$ so $f(a) = a$ is a 1-1 map from M to S.

(ii) Yes. Since every live animal has one heart and one liver, map each animal's heart to its own liver.

(iii) Yes.

$$\begin{array}{ccccccc} 1 & 2 & 3 & 4 & 5 & \ldots \\ | & | & | & | & | & \\ -1 & -2 & -3 & -4 & -5 & \ldots \end{array}$$

4.2.

(i)   (See Theorem 1.8)

$$\begin{array}{ccccccc} 1 & 2 & 3 & 4 & 5 & 6 & \ldots \\ | & | & | & | & | & | & \\ 2 & 3 & 5 & 7 & 11 & 13 & \ldots \end{array}$$

(ii) Order the pairs as follows: Let (m, n) and (p, q) be pairs with $m \leqslant n$ and $p \leqslant q$. If $m + n = p + q$ and $m < p$, put (m, n) before (p, q). If $m + n < p + q$, put (m, n) before (p, q).
      This yields the correspondence:

$$\begin{array}{ccccccccc} 1 & 2 & 3 & 4 & 5 & 6 & 7 & 8 & 9 & \ldots \\ | & | & | & | & | & | & | & | & | & \\ (1,1) & (1,2) & (1,3) & (2,2) & (1,4) & (2,3) & (1,5) & (2,4) & (3,3) & \ldots \end{array}$$

(iii) A direct generalization of (ii). Note that for a given positive integer k, there are only finitely many sets of positive integers whose sum is k.

(iv) This answer could be obtained from (ii) by identifying rational numbers with *ordered* pairs of integers ($\frac{1}{2}$ with $\langle 1, 2 \rangle$; 2 with $\langle 2, 1 \rangle$), but we illustrate another route. Write the rationals in the following matrix, then correlate them with positive integers in the order shown by the arrows, crossing out those which are not expressed in lowest terms:

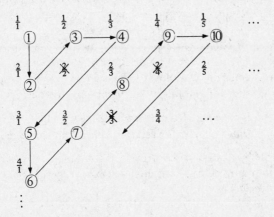

4.3. Assume the hint, and define the one-place real function $g(x) = f_x(x) + 1$. $g(x)$ differs for at least one argument from every other function and hence differs from every other function in the assumed correspondence. Hence, there is no such 1–1 correspondence.

4.4. The barber either shaves himself or he doesn't. If he does not shave himself, then (since he shaves *all* who do not shave themselves) he shaves himself. If he shaves himself, then (since he shaves *only* those who do *not* shave themselves) he does *not* shave himself. He both shaves and does not shave himself. This contradiction shows that he does not exist.

4.5.

(i) Axiom (i) $\quad$ (A $\supset$ (B $\supset$ A))
$$\underline{T_2} \; F_1 \; T_3 \; F_2 \; \underline{F_3}$$

Axiom (ii) ((A $\supset$ (B $\supset$ C)) $\supset$ ((A $\supset$ B) $\supset$ (A $\supset$ C)))
$$\underline{T_5} \; \underline{T_2} \; T_7 \; \underline{F_8} \; F_5 \; F_1 \; T_5 \; T_3 \; T_6 \; F_2 \; T_4 \; F_3 \; F_4$$

Axiom (iii) (($\neg$ B $\supset$ $\neg$ A) $\supset$ (($\neg$ B $\supset$ A) $\supset$ B))
$$T_5 \; F_4 \; T_2 \; T_7 \; \underline{F_8} \; F_1 \; T_5 \; F_4 \; T_3 \; \underline{T_6} \; F_2 \; F_3$$

(ii) Let $P_i, \ldots, P_j$ be the proposition letters occurring in $A \supset B$. Assign to each proposition letter a truth value (T or F). By hypothesis, the resulting value of A is T. Since it is also T for $A \supset B$, by the truth table for $\supset$, it must be T for B.

(iii) Let $B_1, \ldots, B_n$, where $B_n = B$, be a form proof of B. Each member of the sequence is either an axiom or an immediate consequence of preceding members by the rule. The proof is by induction.

  Basis: B_1 is an axiom, hence a tautology, by (i).
  Induction step: assume that for all B_i with $i < k$, B_i is a tautology. If B_k is an axiom, it is a tautology. If B_k is an immediate consequence of B_j and B_h ($= B_j \supset B_k$) for j, $h < k$, then B_k is a tautology by (ii).

(iv) Suppose there is a formula C such that both $\vdash_E C$ and $\vdash_E \neg C$. By (iii), both C and $\neg$C must be tautologies. Since this is impossible, **E** is consistent.

(v) It is decidable by truth tables whether A is a tautology. If A is a tautology, by our assumption, $\vdash_E A$. If A is not a tautology, by (iii), not $\vdash_E A$. Hence, **E** is decidable.

4.6.

(i) (a) $2^9 \cdot 3^{29} \cdot 5^{11}$.

  (b) $2^{11} \cdot 3^{29} \cdot 5^{25} \cdot 7^{29} \cdot 11^{15} \cdot 13^{23} \cdot 17^{27}$.

(ii) No, by the uniqueness of prime factorization (Theorem 1.6, p. 7).

(iii) No; *symbols* have *odd gns*, while *expressions* have *even gns*.

(iv) (a) 3   (b) 2^3.

4.7.

(i) No, again by the prime factorization theorem.

(ii) No. In the *gns* of *expressions*, 2 has an *odd* power, while in the *gns* of *sequences of expressions* 2 has an *even* power:

$$gn(e_0, \ldots, e_m) = 2^{gn(e_0)} \cdot \ldots \cdot p_m^{gn(e_m)}$$

and $$gn(e_0) = 2^{gn(s_0)} \cdot \ldots \cdot p_n^{gn(e_m)}.$$

(iii) No. 2, 4, and 6 are not.

4.8.

(i) $864 = 2^5 \cdot 3^3 = gn(\& \supset)$. The *expression* is '$\& \supset$'.

(ii) 865 is the *gn* of a *variable*, as a *symbol*.

(iii) $866 = 2^1 \cdot 433^1$. This is not a *gn*.

Chapter 5

5.1. Only $=$ is a symbol of our formal theory. Although $<, \sim, 2, \neq, \{, |$ may be used in abbreviations or to make reading easier, they do *not* belong to the formal theory.

5.2. (i), (ii), (iii), (iv), (v), and (vii) are expressions according to the definition. An expression is no more than a finite string of symbols of the formal theory. However (viii), (ix), and (x) are *not* expressions. Expression (vii) has five occurrences of symbols. It is not the same expression as $0\forall0 =$.

5.3. Only (i), (iii), (iv), (v), and (ix) are formal *terms*. (vi), (vii), (xiii), (xiv), (xvi), and (xvii) are not even *expressions*. (ii), (x), and (xi) break bracket rules. (xii) and (xv) are *formulas*, not *terms*. Nothing in the rules allows (viii) as a *term*. (Note: we might 'express' 3! by $(0''' \cdot (0'' \cdot 0'))$ and 2^2 by $(0'' \cdot 0'')$, but it is not at all obvious how we might express $a!$ or 2^a, where a is a variable.)

5.4. Only (ii), (iii), (x), (xii), (xiii), (xv), (xvi), and (xvii) are *formulas*. The examples illustrate various points including the following. (iii) and (xii) draw out the fact that truth or falsity under the intended interpretation has no bearing on whether an *expression* is a formula. (iv) is used as an abbreviation, to make reading formulas easier, for

$$\exists c((c' + a) = b),$$

but is not itself a *formula*. (v) is an abbreviation for

$$(0' < a \,\&\, \neg\exists b\exists c(0' < b \,\&\, (b < a \,\&\, ((b \cdot c) = a))))$$

and will be read 'a is prime'. (viii) notes that our formal theory lacks proposition letters. With respect to (xviii) and (xix), on the face of it, we do not yet have a way of expressing the power function in **N**. We shall return to this in a later chapter.

5.5.

(i) $((a + 0') = a)$

$$\underline{\quad}'$$

$$- + -$$

$$\overline{\qquad\quad} = -$$

(ii) $\forall a \exists a(0 = b)$

$$
\begin{array}{l}
\quad\quad - = - \\
\exists a \underline{} \\
\forall a \underline{}
\end{array}
$$

(iii) $\forall a \exists b \exists q \exists r(a = b \cdot q + r)$

$$
\begin{array}{l}
\quad\quad\quad\quad\quad - \cdot - \\
\quad\quad\quad\quad - + - \\
\quad\quad\quad - = \underline{} \\
\quad\quad \exists r \underline{} \\
\quad \exists q \underline{} \\
\exists b \underline{} \\
\forall a \underline{}
\end{array}
$$

(iv) $\exists c(\exists c(c' + a = b) \supset \neg a = b + c)$

$$
\begin{array}{l}
\quad\quad -' \\
\quad\quad - + - \quad\quad\quad\quad - + - \\
\quad\quad \underline{} = -; \quad - = \underline{} \\
\quad \exists c \underline{}; \ \neg \underline{} \\
\exists c \underline{}
\end{array}
$$

(v) $(A \supset B) \supset ((A \supset \neg B) \supset \neg A)$

$$
\begin{array}{l}
\quad\quad\quad\quad\quad \neg - \quad \neg - \\
\quad - \supset -; \quad - \supset - \\
\quad\quad\quad\quad \underline{} \supset \underline{} \\
\underline{} \supset \underline{}
\end{array}
$$

5.6.

(i) In α, all occurrences of b and the first occurrence of a are free; all other occurrences are bound. In β, all occurrences of a and b are free; all occurrences of c are bound.

(ii) In α, the first occurrence of a is free in all terms and subformulas in which it occurs; the second occurrence is bound throughout. The third is free in a, a', $a' + r$, $a' + r = b$ and bound in all the other subformulas in which it occurs. In β, all occurrences of a are free in all terms and subformulas in which they occur.

(iii) In α, only a is both a free and a bound variable of the formula. In β, no variable is both.

(iv) In α, all occurrences of q and r are bound by the quantifiers $\exists q$ and $\exists r$; the second and third occurrences of a are bound by $\exists a$. In β, each occurrence of c is bound either by the quantifier in which it occurs or by its immediately preceding quantifier.

5.7. (i) $\exists c(0 \cdot c = b)$, (ii) $\exists c(b \cdot c = b)$, (iii) $\exists c(c \cdot c = b)$.
    Note that substitution must always be for the original variable in the original formula. One cannot, in this example, get $A(c)$ from $A(b)$ by substituting c for b in $\exists c(b \cdot c = b)$.

5.8.

(i) Only $(b + c')$ is *not* free for a in $\exists c(c' + a = b)$.

(ii) Only 0 is free for b. All are free for r. $a + r$ is, trivially, free for r in $\neg b = 0 \supset \exists q \exists r(a = b \cdot q + r \& \exists a(a' + r = b))$ because r is not free in this formula. So, $A(a + r)$ is the same as $A(r)$.

(iii) Only $a \cdot c$ is *not* free for a. Only $a \cdot c$ is *not* free for b. All are (trivially) free for c, since c is not free anywhere in $\exists c(c' + 0' = a) \supset \neg(\exists c(a \cdot c = b)$ & $\exists c(a \cdot c = b'))$.

5.9. (1) Axiom 16; (2) Axiom schema 1(a); (3) Axiom schema 1(a); (4) Rule 2, 1, 3; (5) Rule 9, 4; (6) Rule 9, 5; (7) Rule 9, 6; (8) Rule 2, 2, 7; (9) Axiom schema 10; (10) Rule 2, 8, 9; (11) Axiom schema 10; (12) Rule 2, 10, 11; (13) Axiom schema 10; (14) Rule 2, 12, 13; (15) Axiom 18; (16) Rule 2, 15, 14; (17) Rule 2, 15, 16.

5.10. (1) Axiom schema 1(a); (2) Axiom schema 1(b); (3) Rule 2, 1, 2; (4) Axiom schema 1(a); (5) Rule 2, 4, 3.

5.11.

(i) 1, 3, and 4 *depend on* assumption 1. Variables are *not* held constant, because line 4 is obtained by applying Rule 9 to a formula which depends on an assumption formula with x free.

(ii) 1, 3, 4, and 5 *depend on* assumption 1. Variables are *not* held constant, because line 5 is obtained by applying Rule 9 to a formula which depends on assumption formula with x free. Note: the application of Rule 12 does *not* violate the conditions for holding variables constant.

(iii) 1, 3, 7, 9, and 10 *depend on* assumption 1. Variables are *not* held constant, because line 10 is obtained by applying Rule 12 to a formula which depends on an assumption formula with x free. Although line 10 is provable in **N**, *this* sequence does not prove it.

5.12. *Rule 12* $B_k (= \exists x D(x) \supset C)$ is an immediate consequence of a preceding formula B_m, $(D(x) \supset C)$, where $m < k$, by Rule 12. The variables are held constant. There are two possible cases: either (i) B_m depends on A; or (ii) B_m does not depend on A.

Case (i) If B_m depends on A, then (because variables are held constant) A cannot contain x free. Since C does not contain x free, neither does $(A \supset C)$. By the induction hypothesis, $\Gamma \vdash A \supset (D(x) \supset C)$, so by (*)$A \supset (B \supset C) \vdash B \supset (A \supset C)$, we get $\Gamma \vdash D(x) \supset (A \supset C)$. Thus, by Rule 12, $\Gamma \vdash \exists x D(x) \supset (A \supset C)$. So, again by (*), $\Gamma \vdash A \supset (\exists x D(x) \supset C)$.

Case (ii) If B_m does not depend on A, neither does $B_k(\exists x D(x) \supset C)$. Hence, there is a sequence establishing $\Gamma \vdash \exists x D(x) \supset C$. Add to this $\vdash (\exists x D(x) \supset C) \supset (A \supset (\exists x D(x) \supset C))$, by Axiom 1(a). Apply Rule 2, and we have $\Gamma \vdash A \supset (\exists x D(x) \supset C)$.

5.13. By repeated applications of Axiom 3

$$A_1, \ldots, A_m \vdash (B_1 \ \& \ \ldots \ \& \ B_p).$$

By repeated applications of the Deduction Theorem,

$$\vdash A_1 \supset (A_2 \supset (A_3 \supset \ldots (A_m \supset (B_1 \ \& \ \ldots \ \& \ B_p) \ldots))).$$

By repeated applications of $\vdash (A \supset (B \supset C)) \supset ((A \ \& \ B) \supset C)$,

$$\vdash (A_1 \ \& \ \ldots \ \& \ A_m) \supset (B_1 \ \& \ \ldots \ \& \ B_p).$$

Similarly, $\vdash (B_1 \ \& \ \ldots \ \& \ B_p) \supset C$. By Axioms 1(a) and 1(b), $\vdash (A_1 \ \& \ \ldots \ \& \ A_m) \supset C$. Therefore, $A_1, \ldots, A_m \vdash C$.

5.14. $\supset$-intro. is the Deduction Theorem, which we have already proved.

$\supset$-elim.	1.	A	assumption.
	2.	$A \supset B$	assumption.
	3.	B	1, 2, *modus ponens.*
&-intro.	1.	A	assumption.
	2.	B	assumption.
	3.	$A \supset (B \supset (A \ \& \ B)$	Axiom 3.
	4.	$B \supset (A \ \& \ B)$	1, 3, *modus ponens.*
	5.	$A \ \& \ B$	2, 4, *modus ponens.*

&-elim.   1.   A & B   assumption.

             2.   (A & B) ⊃ A   Axiom 4(a) (4(b) for B).

             3.   A   1, 2, *modus ponens.*

∨-intro.   1.   A   assumption.

             2.   A ⊃ (A ∨ B)   Axiom 5(a) (5(b) for B).

             3.   A ∨ B   1, 2, *modus ponens.*

∨-elim.   By the Deduction Theorem, Γ ⊢ A ⊃ C and Γ ⊢ B ⊃ C. Clearly
             A ⊃ C, B ⊃ C ⊢ (A ∨ B) ⊃ C by Axiom 6. It follows that
             Γ, A ∨ B ⊢ C by Theorem 5.3.

¬-intro.   By the Deduction Theorem, Γ ⊢ A ⊃ B and Γ ⊢ A ⊃ ¬B. Clearly
             A ⊃ B, A ⊃ ¬B ⊢ ¬A, by Axiom 7. Hence, Γ⊢¬A by Theorem
             5.3.

¬-elim.   1.   ¬¬A   assumption.

             2.   ¬¬A ⊃ A   Axiom 8.

             3.   A   1, 2, *modus ponens.*

∀-intro.   If Γ ⊢ A(x) and Γ does not contain a formula with x free, A(x)
             does not depend on a formula with x free. Let C be an axiom not
             containing x free, and add to the sequence which shows Γ ⊢ A(x):

                  A(x) ⊃ (C ⊃ A(x))   Axiom 1(a).

                  C ⊃ A(x)   1, 2, *modus ponens.*

                  C ⊃ ∀xA(x)   3, Rule 9.

                  C   Axiom.

                  ∀xA(x)   4, 5, *modus ponens.*

∀-elim.   1.   ∀xA(x) ⊃ A(t)   Axiom 10.

             2.   ∀xA(x)   assumption.

             3.   A(t)   1, 2, *modus ponens.*

∃-intro.   1.   A(t)   assumption.

             2.   A(t) ⊃ ∃xA(x)   Axiom 11.

             3.   ∃xA(x)   1, 2, *modus ponens.*

∃-elim.   By the Deduction Theorem, Γ ⊢ A(x) ⊃ C. Since C does not
             contain x free and A(x) ⊃ C does not depend on a formula with
             x free, Rule 12 yields Γ ⊢ ∃xA(x) ⊃ C. Hence, Γ, ∃xA(x) ⊢ C.

Chapter 6

Although the routes to formal theorems which we outline here are correct, there are of course innumerable alternative solutions.

6.1.   The proof is by formal induction, $A(c)$ being $a = b \supset ac = bc$.

1.	$\vdash a0 = 0$	Axiom 20.
2.	$\vdash 0 = b0$	Axiom 20, *101.
3.	$\vdash a0 = b0$	1, 2, &-intro., *102.
4.	$\vdash a = b \supset a0 = b0$	3, Axiom 1(a).
5.	$ac = bc \vdash ac' = ac + a$	Axiom 21, Lemma 5.1 (ii).
6.	$ac = bc \vdash ac + a = bc + a$	*104.
7.	$ac = bc \vdash ac' = bc + a$	5, 6 as in steps 1–5 of *104.
8.	$a = b \vdash bc + a = bc + b$	*105.
9.	$\vdash bc + b = bc'$	Axiom 21, *101.
10. $a = b, ac = bc \vdash ac' = bc'$		7–9, as in steps 1–5 of *104 (twice).
11.	$\vdash ac = bc \supset (a = b \supset ac' = bc')$	10, $\supset$-intro.
12.	$\vdash (a = b \supset ac = bc) \supset (a = b \supset ac' = bc')$	11 by $\vdash (B \supset (A \supset C)) \supset ((A \supset B) \supset (A \supset C))$.
13.	$\vdash a = b \supset ac = bc$	4, 12, $\forall$-intro., &-intro. then formal induction, Axiom 13.

6.2. By the *depth* of r we shall mean the number of operators within whose scope r (i.e., the specified occurrence of r) lies. For instance, in $a \cdot b' + a$, the depth of the second a is 1, the depth of the first a is $2(+, \cdot)$, and the depth of b is $3(+, \cdot, ')$. Then the proof is by induction on the *depth* d of r in t_r.

Basis: r is at depth 0. Then, t_r is r_1, t_s is s, and $r = s \vdash t_r = t_s$ is immediate.

Induction step: r is at depth $d + 1$ in t_r. By the induction hypothesis, $r = s \vdash u_r = u_s$ for any term u_r in which the specified occurrence of r lies at depth d. Now t_r must have one of the forms $(u_r)'$, $u_r + w$, $w + u_r$, $u_r \cdot w$, $w \cdot u_r$ where u_r and w are terms and r is at depth d in u_r. Hence, by the appropriate lemma from among *110–*114, $u_r = u_s \vdash t_r = t_s$. But we know by the induction hypothesis that $r = s \vdash u_r = u_s$, so, by Theorem 5.3, $r = s \vdash t_r = t_s$. This completes the induction.

6.3.   Assume $\Gamma \vdash r = s$ and $\Gamma \vdash s = t$. By *115, $r = s$, $s = t \vdash r = t$. So, by Theorem 5.3, if $\Gamma \vdash r = s$ and $\Gamma \vdash s = t$, then $\Gamma \vdash r = t$.

6.4.   *117. Proof by induction on c.

1. $\vdash (a + b) + 0 = a + b$ Axiom 18 (substituting $a + b$ for a).

2. $b + 0 = b \vdash a + (b + 0) = a + b$ Replacement Theorem, $r = b + 0$, $s = b$.

3. $\vdash a + b = a + (b + 0)$ 2, Axiom 18 (substituting b for a), $\vdash r = s \sim s = r$ and *modus ponens*.

4. $\vdash (a + b) + 0 = a + (b + 0)$
                    1, 3, and Theorem 6.3.

5. $\vdash (a + b) + c' = ((a + b) + c)'$ Axiom 19 (substituting $a + b$ for a and c for b).

6. $(a + b) + c = a + (b + c) \vdash ((a + b) + c)' = (a + (b + c))'$ Replacement Theorem $r = (a + b) + c$, $s = a + (b + c)$.

7. $\vdash (a + (b + c))' = a + (b + c)'$ Substitution in Axiom 19 and $\vdash r = s \sim s = r$.

8. $(b + c)' = b + c' \vdash a + (b + c)' = a + (b + c')$
                    Replacement Theorem.

9. $\vdash a + (b + c)' = a + (b + c')$ 8, ⊃-intro. Axiom 19 (substituting b for a and c for b), and *modus ponens*.

10. $(a + b) + c = a + (b + c) \vdash (a + b) + c' = a + (b + c')$ 5, 6, 7, 9, and repeated application of Theorem 6.3.

11. $\vdash (a + b) + c = a + (b + c)$ 1 and 10 using Axiom 13, A(c) being $(a + b) + c = a + (b + c)$. The steps are as in the proof of $\vdash 0 + a = a$, lines 4–8.

*118   Proof by induction on b.

1.	$a + 0 = a \vdash (a + 0)' = a'$	Replacement Theorem.
2.	$\vdash (a + 0)' = a'$	1, Axiom 18, $\supset$-intro., and *modus ponens.*
3.	$\vdash a' + 0 = a'$	Axiom 18 (substituting a' for a).
4.	$\vdash a' + 0 = (a + 0)'$	Theorem 6.3 and properties of identity.
5.	$\vdash a' + b' = (a' + b)'$	Axiom 19 (substituting a' for a).
6.	$a' + b = (a + b)' \vdash (a' + b)' = ((a + b)')'$	Replacement Theorem $r = a' + b, s = (a + b)'$.
7.	$(a + b)' = a + b' \vdash ((a + b)')' = (a + b')'$	Replacement Theorem $r = (a + b)', s = a + b'$.
8.	$\vdash ((a + b)')' = (a + b')'$	7, Axiom 19, $\supset$-intro., *modus ponens,* and Theorem 6.3.
9.	$a' + b = (a + b)' \vdash a' + b' = (a + b')'$	5, 6, 8, and Theorem 6.3 repeatedly.
10.	$\vdash a' + b = (a + b)'$	4, 9, $\supset$-intro., $\forall$-intro., &-intro., Axiom 13 (A(b) being $a' + b = (a + b)')$ and *modus ponens.*

*119   Proof by induction on a

1.	$\vdash b + 0 = b$	substituting b for a in Axiom 18.
2.	$\vdash 0 + b = b$	substituting b for a in $\vdash 0 + a = a.$
3.	$\vdash 0 + b = b + 0$	Theorem 6.3.
4.	$\vdash a' + b = (a + b)'$	*118.
5.	$a + 0 = b + a \vdash (a + b)' = (b + a)'$	Replacement Theorem $r = a + b, s = b + a$.
6.	$\vdash (b + a)' = b + a'$	Axiom 19 (substituted), and identity.

7. $a + b = b + a \vdash a' + b = b + a'$ 4, 5, 6, and Theorem 6.3, repeatedly.

8. $\vdash a + b = b + a$ 3, 7, $\supset$-intro., $\forall$-intro., &-intro., Axiom 13 ($A(a)$ being $a + b = b + a$) and *modus ponens.*

*120   Proof by induction on c.

1. $\vdash a(b + 0) = ab + a \cdot 0$ Axiom 18, Axiom 20, Replacement Theorem, and Theorem 6.3.

2. $b + c' = (b + c)' \vdash a(b + c') = a(b + c)'$ Replacement Theorem $r = b + c', s = (b + c)'$.

3. $\vdash a(b + c') = a(b + c)'$ 2 and Axiom 19.

4. $\vdash a(b + c)' = a(b + c) + a$ Axiom 21 (substituting $b + c$ for b)

5. $a(b + c) = ab + ac \vdash a(b + c) + a = (ab + ac) + a$ Replacement Theorem.

6. $\vdash (ab + ac) + a = ab + (ac + a)$ *117.

7. $ac + a = ac' \vdash ab + (ac + a) = ab + ac'$ Replacement Theorem.

8. $a(b + c) = ab + ac \vdash a(b + c') = ab + ac'$ 3, 4, 5, 6, 7, Axiom 21 (substituting for b), *modus ponens* and Theorem 6.3, repeatedly.

9. $\vdash a(b + c) = ab + ac$ 1, 8, and induction.

*121   Proof by induction on c.

1. $\vdash (ab) \cdot 0 = a(b \cdot 0)$ Axiom 20, Replacement Theorem and Theorem 6.3.

2. $\vdash (ab)c' = (ab)c + ab$ Axiom 21 (substituting ab for a and b for c).

3. $(ab)c = a(bc) \vdash (ab)c + ab = a(bc) + ab$ Repeat last Theorem, $r = (ab)c, s = a(bc)$.

4. $\vdash a(bc) + ab = a(bc + b)$ *120.

5. $\vdash a(bc + b) = a(bc')$ Axiom 21, Replacement Theorem.

6. $(ab)c = a(bc) \vdash (ab)c' = a(bc')$ 2, 3, 4, 5, and Theorem
                                               6.3, repeatedly.

7. $\vdash (ab)c = a(bc)$ 1, 6, and induction.

*122 Proof by induction on b.

1. $\vdash a' \cdot 0 = a \cdot 0 + 0$ Axiom 20, Axiom 18,
                                                             Replacement Theorem,
                                                             and Theorem 6.3.

2. $\vdash a'b' = a'b + a'$ Axiom 21 (substituted).

3. $a'b = ab + b \vdash a'b + a' = (ab + b) + a'$ Replacement Theorem.

*4. $\vdash \quad = ((ab + b) + a)'$ Axiom 19.

5. $\vdash \quad = (ab + (b + a))'$ *117.

6. $\vdash \quad = (ab + (a + b))'$ *119.

7. $\vdash \quad = ((ab + a) + b)'$ *117 and Replacement
                                                        Theorem.

8. $\vdash \quad = (ab' + b)'$ Axiom 21 and Replace-
                                                    ment Theorem.

9. $\vdash \quad = ab' + b'$ Axiom 19.

10. $a'b = ab + b \vdash a'b' = ab' + b'$ 2, 3, 4, 5, 6, 7, 8, 9, and
                                             Theorem 6.3, repeatedly.

11. $\vdash a'b = ab + b$ 1, 10, and induction.

* Where the l.h.s. of an equality is omitted it is simply the r.h.s. of the previous equality.

To prove $\vdash 0 \cdot a = 0$. Proof by induction on a.

1. $\vdash 0 \cdot 0 = 0$ Axiom 20.

2. $\vdash 0 \cdot a' = 0 \cdot a + 0$ Axiom 21.

3. $0 \cdot a = 0 \vdash 0 \cdot a + 0 = 0 + 0$ Replacement Theorem.

4. $\vdash \quad = 0$ Axiom 18.

5. $0 \cdot a = 0 \vdash 0 \cdot a' = 0$ 2, 3, 4, and Theorem 6.3.

6. $\vdash 0 \cdot a = 0$ 1, 5, and induction.

*123 Proof by induction on b

1. $\vdash a \cdot 0 = 0 \cdot a$ Axiom 20, $\vdash 0 \cdot a = 0$,
                                             and Theorem 6.3.

2. $\quad\quad\quad \vdash ab' = ab + a$ — Axiom 21.

3. $\quad ab = ba \vdash ab + a = ba + a$ — Replacement Theorem.

4. $\quad\quad\quad \vdash \quad\quad = b'a$ — *122.

5. $\quad ab = ba \vdash ab' = b'a$ — 2, 3, 4, and Theorem 6.3.

6. $\quad\quad\quad \vdash ab = ba$ — 1, 5, and induction.

*127

1. $\quad\quad\quad \vdash a \cdot 0' = a \cdot 0 + a$ — Axiom 21.

2. $\quad a \cdot 0 = 0 \vdash a \cdot 0 + a = 0 + a$ — Replacement Theorem.

3. $\quad\quad\quad \vdash a \cdot 0' = 0 + a$ — 1, 2, Axiom 20 and Theorem 6.3.

4. $\quad\quad\quad \vdash a \cdot 0' = a$ — $\vdash 0 + a = a$ and Theorem 6.3.

*132 Proof by induction on c.

1. $\quad\quad\quad\quad \vdash a = a + 0$ — Axiom 18, *100, and *modus ponens.*

2. $\quad a + 0 = b + 0 \vdash a + 0 = b + 0$ — Lemma 5.1.

3. $\quad\quad\quad\quad \vdash b + 0 = b$ — Axiom 18.

4. $\quad\quad\quad\quad \vdash a + 0 = b + 0 \supset a = b$ — 1, 2, 3, Theorem 6.3. (or *115), and $\supset$-intro.

5. $\quad\quad\quad\quad \vdash (a + c)' = (b + c)' \supset a + c = b + c$ — Axiom 14 (substituted).

6. $\quad a + c' = b + c' \vdash (a + c)' = (b + c)'$ — Lemma 5, Axiom 19 and Theorem 6.3 (or *115).

7. $\quad a + c' = b + c' \vdash a + c = b + c$ — 5, 6, if $\vdash A \supset B$ then $A \vdash B$, and Theorem 5.3.

8. $\quad a + c = b + c \supset a = b, a + c' = b + c' \vdash a = b$ — Theorem 5.3, 7, and *modus ponens.*

9. $\quad a + c = b + c \supset a = b \vdash a + c' = b + c' \supset a = b$ — 8 and $\supset$-intro.

10. $\quad\quad\quad\quad \vdash a + c = b + c \supset a = b$ — 4, 9, and induction.

6.6.   *153. Immediate from *127, and Axiom 11.

*154

1.	$am = b, bn = c \vdash a(mn = c$	*121, Replacement Theorem, and Theorem 6.3.
2.	$a(mn) = c \vdash \exists z(az = c)$	$\exists$-intro.
3.	$am = b, bn = c \vdash \exists z(az = c)$	1, 2, and Theorem 5.3.
4.	$\exists x(ax = b), \exists y(by = c) \vdash \exists z(az = c)$	$\exists$-elim. twice.
5.	$\exists x(ax = b) \,\&\, \exists y(by = c) \vdash \exists x(ax = b)$	&-elim.
6.	$\exists x(ax = b) \,\&\, \exists y(by = c) \vdash \exists y(by = c)$	&-elim.
7.	$\exists x(ax = b) \,\&\, \exists y(by = c) \vdash \exists z(az = c)$	4, 5, 6, Theorem 5.3.
8.	$\vdash \exists x(ax = b) \,\&\, \exists y(by = c) \supset \exists z(az = c)$	$\supset$-intro.

*155

1.	$1 + p' = a \vdash a > 1$	$\exists$-intro.
2.	$\vdash a > 1 \supset (n < m \sim an < am)$	*145(a).
3.	$1 + p' = a \vdash n < m \sim an < am$	1, 2, if $\vdash A \supset B$ then $A \vdash B$, and Theorem 5.3.
4.	$\vdash b < b'$	*135(a).
5.	$1 + p' = a, an = b, am = b' \vdash an < am$	4, Replacement Theorem.
6.	$1 + p' = a, an = b, am = b' \vdash n < m$	3, 5, Theorem 5.3, and *modus ponens.*
7.	$n + c' = m, an = b, am = b' \vdash b' = am$	Lemma 5.1(i), (ii).
8.	$n + c' = m, an = b, am = b' \vdash \quad = an + ac'$	7, Replacement Theorem and *120.
9.	$n + c' = m, an = b, am = b' \vdash \quad = b + ac + a$	Axiom 21, Replacement Theorem.
10.	$1 + p' = a, an = b, am = b' \vdash \quad = b' + ac + p'$	Replacement Theorem, *119, and Axiom 19, etc.
11.	$1 + p' = a, an = b, am = b' \vdash \quad = b' + (ac + p)'$	Axiom 19 and Replacement Theorem.
12.	$1 + p' = a, an = b, am = b' \vdash b' < b'$	6, 7, 8, 9, 10, 11, *115, $\exists$-intro., $\exists$-elim. and 6.
13.	$\vdash \neg b' < b'$	*140.

14.	$a > 1, a|b, a|b' \vdash b' < b'$	12 and $\exists$-elim.
15.	$a > 1, a|b \vdash \neg a|b'$	13, 14, and $\neg$-intro.
16.	$a > 1 \vdash \neg(a|b \,\&\, a|b')$	15, tautology.
17.	$\vdash a > 1 \supset \neg(a|b \,\&\, a|b')$	16 and $\supset$-intro.

6.7. Let $B(x)$ be $\forall y(y \leqslant x \supset A(y))$ and C be $\forall x(\forall y(y < x \supset A(y)) \supset A(x))$

1.	$C \vdash \forall y(y < 0 \supset A(y)) \supset A(0)$	$\forall$-elim.
2.	$\vdash \neg y < 0$	
3.	$\vdash \forall y(y < 0 \supset A(y))$	2, tautology, and $\forall$-intro.
4.	$C \vdash A(0)$	1, 3, and *modus ponens*.
5.	$C \vdash B(0)$	3, 4, and $\vdash (y < k \supset A(y))$ $\&\ A(k) \supset (y \leqslant k \supset A(y))$.
6.	$C, B(x) \vdash \forall y(y < x' \supset A(y))$	$B(x)$ and $\vdash y \leqslant x \sim y < x'$.
7.	$C \vdash \forall y(y < x' \supset A(y)) \supset A(x')$	$\forall$-elim.
8.	$C, B(x) \vdash A(x')$	6, 7, and *modus ponens*.
9.	$\vdash y \leqslant x' \supset y < x' \lor y = x'$	definition and tautology.
10.	$C, B(x) \vdash y < x' \supset A(y)$	6 and $\forall$-elim.
11.	$C, B(x) \vdash y = x' \supset A(y)$	8 and Replacement Theorem.
12.	$C, B(x) \vdash B(x')$	9, 10, 11, tautology and $\forall$-intro.
13.	$C \vdash B(x)$	5, 12, and induction.
14.	$C \vdash x \leqslant x \supset A(x)$	13 and $\forall$-elim.
15.	$\vdash x \leqslant x$	*100 and definition.
16.	$C \vdash \forall x(A(x))$	14, 15, *modus ponens*, and $\forall$-intro.
17.	$\vdash C \supset \forall x(A(x))$	$\supset$-intro.

6.8. Yes. All are provable.

(i) 1.	$b \neq 0, a = b \vdash a|a$	*153.
2.	$\vdash a|b$	1, definition, Replacement Theorem.
3.	$\vdash d > a \supset \neg d|a$	*156, *141, $\vdash d > a \supset \neg d = a$.
4.	$\vdash d > a \supset \neg d|b$	3, Replacement Theorem.

5. $\vdash \neg(d > a \,\&\, d|a \,\&\, d|b)$ 3, 4, tautology.

6. $\vdash \neg\exists d(d > a \,\&\, d|a \,\&\, d|b)$ 5, $\forall$-intro.,
                                            $\vdash \forall d \neg Fd \sim \neg\exists d Fd$.

7. $\vdash g(a, b) = a$ 1, 2, 6, def.

(ii)   1. $a + c' = b$, $g(a, b) = j$,

             $g(a, c') = k \vdash j|a \,\&\, j|b \,\&\, \neg \exists m(m > j \,\&\, m|a \,\&\, m|b)$

2. $\vdash k|a \,\&\, k|c' \,\&\, \neg\exists n(n > k \,\&\, n|a \,\&\, n|c')$

3. $\vdash j|a + c'$ 1, Replacement Theorem.

4. $\vdash j|c'$ 3, $\vdash j|a \,\&\, j|a + c' \supset j|c'$.

5. $\vdash k|c'$ 2, &-elim.

6. $\vdash \neg j > k$ 4, 5, 2,
                                              $(\neg\exists n(n > k \,\&\, n|c'))$.

7. $\vdash \neg k > j$ 1, 2, $(\neg\exists m(m > j \,\&\, m|a))$.

8. $\vdash j = k$ 6, 7, *139.

(iii)   Similar to (ii).

Chapter 7

7.1.

*85   1. $\neg A(x)$, $\forall x A(x) \vdash A(x)$ $\forall$-elim.

2. $\neg A(x)$, $\forall x A(x) \vdash \neg A(x)$

3. $\neg A(x) \vdash \neg \forall x A(x)$ 1, 2, $\neg$-intro.

4. $\exists x \neg A(x) \vdash \neg \exists x A(x)$ 3, $\exists$-elim.

5. $\vdash \exists x \neg A(x) \supset \neg \forall X A(x)$ 4, $\supset$-intro.

---

6. $\neg\exists x \neg A(x)$, $\neg A(x) \vdash \exists x \neg A(x)$ $\exists$-intro.

7. $\neg\exists x \neg A(x)$, $\neg A(x) \vdash \neg\exists x \neg A(x)$

8. $\neg\exists x \neg A(x) \vdash \neg\neg A(x)$ 6, 7, $\neg$-intro.

9. $\neg\exists x \neg A(x) \vdash A(x)$ 8, $\neg$-elim.

10. $\neg\exists x \neg A(x) \vdash \forall x A(x)$ 9, $\forall$-intro.

11. $\vdash \neg\exists x \neg A(x) \supset \forall x A(x)$ 10, $\supset$-intro.

12. $\vdash \neg\forall x A(x) \supset \exists x \neg A(x)$ 11, $\vdash (A \supset B) \supset (\neg B \supset \neg A)$.

*98. 1. $A(x) \supset B, \forall x A(x) \vdash A(x)$ $\forall$-elim.

2. $A(x) \supset B, \forall x A(x) \vdash A(x) \supset B$

3. $A(x) \supset B, \forall x A(x) \vdash B$ 1, 2, $\supset$-elim.

4. $\quad\quad\quad A(x) \supset B \vdash \forall x A(x) \supset B$ 3, $\supset$-intro.

5. $\quad\quad \exists x(A(x) \supset B) \vdash \forall x A(x) \supset B$ 4, $\exists$-elim.

---

6. $\quad\quad \forall x A(x) \supset B \vdash \neg \forall x A(x) \lor B$

7. $\quad\quad\quad \neg \forall x A(x) \vdash \exists x \neg A(x)$ *85.

8. $\quad\quad\quad\quad \neg A(x) \vdash A(x) \supset B$ tautology, etc.

9. $\quad\quad\quad\quad \neg A(x) \vdash \exists x(A(x) \supset B)$ 8, $\exists$-intro.

10. $\quad\quad\quad \neg \forall x A(x) \vdash \exists x(A(x) \supset B)$ 7–9, $\exists$-elim.

11. $\quad\quad\quad\quad\quad\quad B \vdash A(x) \supset B$ Axiom 1(a).

12. $\quad\quad\quad\quad\quad\quad B \vdash \exists x(A(x) \supset B)$ 11, $\exists$-intro.

13. $\quad\quad \forall x A(x) \supset B \vdash \exists x(A(x) \supset B)$ 6, 10, 12, $\supset$-intro.

7.2.

1. $\quad a = b \vdash a + 0' = b'$ Axioms 17_D, 18_D, 19_D, and Replacement Theorem.

2. $\quad a = b \vdash b + 0' = a'$ as above in 1.

3. $\quad\quad \vdash a = b \supset a < b' \;\&\; b < a'$ 2, 3, $\exists$-intro., &-intro., and $\supset$-intro.

4. $\neg a = b \vdash a < b \lor b < a$ *139_D and tautology.

5. $\quad a < b \vdash \neg b < a'$ $a + c' = b$, $b + d' = a' \vdash (c + d)' = 0$, and *$132_D$.

6. $b < a \vdash \neg a < b'$ similarly

7. $\quad\quad \vdash \neg a = b \supset \neg(a < b' \;\&\; b < a')$ 4, 5, 6, $\lor$-intro., and $\supset$-intro.

8. $\quad\quad \vdash a < b' \;\&\; b < a' \supset a = b$ 7 and tautology.

7.3.

1. $\quad\quad \vdash \neg a = b \supset a < b \lor b < a$ *139_D and tautology.

2. $\quad a < b \vdash \neg a = b$ $a + c' = b$, $a = b \vdash c' = 0$ and *132_D.

3. $\quad b < a \vdash \neg a = b$ $\qquad$ similarly.

4. $\qquad \vdash a < b \vee b < a \supset \neg a = b$ $\qquad$ 2, 3, $\vee$-elim., and $\supset$-intro.

7.4.

1. $\quad \neg a < b \vdash a = b \vee b < a$ $\qquad$ *139$_D$ and tautology.

2. $\quad a = b \vdash b < a'$ $\qquad$ from Exercise 7.2, above.

3. $\quad b < a \vdash b < a'$ $\qquad$ *134(a), *135(a), etc.

4. $\qquad \vdash \neg a < b \supset b < a'$ $\qquad$ 1, 2, 3, $\vee$-elim., and $\supset$-intro.

5. $\qquad \vdash b < a' \supset \neg a < b$ $\qquad$ $b + c' = a'$, $a + d' = b \vdash (c + d)' = 0$, and *132$_D$.

7.5.

1. $\qquad \vdash \neg a \equiv b(\bmod n) \supset a \equiv b + 1(\bmod n) \vee \ldots \vee a \equiv b + n - 1(\bmod n)$
$\qquad$ (Theorem 7.6 (vii) and $\vdash P \vee Q \sim (\neg P \supset Q)$.

2. $\qquad a \equiv b + 1(\bmod n), a \equiv b(\bmod n) \vdash 1 \equiv 0(\bmod n)$
$\qquad$ Theorem 7.6 (ii), (iii), and (iv).

3. $\qquad \vdash \neg 1 \equiv 0(\bmod n)$
$\qquad$ Theorem 7.6 (viii), $\vdash \neg(1 = 0 \vee 1 \geqslant n)$, and tautology.

4. $\qquad a \equiv b + 1(\bmod n) \vdash \neg a \equiv b(\bmod n)$
$\qquad$ 2, 3, and $\neg$-intro.

$\vdots$ $\qquad\qquad$ $\vdots$

$1 + 3(n-1).$ $\qquad a \equiv b + n - 1(\bmod n) \vdash \neg a \equiv b(\bmod n)$
$\qquad$ similarly

$\qquad \vdash a \equiv b + 1(\bmod n) \vee \ldots \vee a \equiv b + n - 1(\bmod n) \supset \neg a \equiv b(\bmod n).$

$\qquad$ 2,..., $1 + 3(n - 1)$, and proof by cases.

7.6. Use the following method: if s$'$ is a 'subterm' of t which contains x then s$'$ is replaced by s $+ 0'$, e.g.

$$b + (x' + c')' = b + (x' + c') + 0' = b + (x + 0' + c') + 0'.$$

Repeat this until there is no successed subterm containing x. Now gather the xs to the left by the associativity and commutativity of +, e.g.,

$$x + (c + x) = x + (x + c) = (x + x) + c.$$

By the properties of '=' this easily yields $\vdash t = x \cdot n + r$.

7.9.

1. $x \cdot p \equiv t(\bmod n), t \equiv 0(\bmod n) \vdash t \equiv 0(\bmod n) \ \& \ x \cdot p \equiv 0$
$(\bmod n)$

Theorem 7.6 (iii) and &-intro.

2. $x \cdot p \equiv t(\bmod n), t \equiv 0(\bmod n) \vdash \text{r.h.s.}$ 1 and ∨-intro.

⋮   ⋮

2n. $x \cdot p \equiv t(\bmod n), t \equiv n - 1(\bmod n) \vdash \text{r.h.s.}$ similarly.

2n + 1. $x \cdot p \equiv t(\bmod n) \vdash \text{r.h.s.}$ ∨-elim. and
Theorem 7.6 (vi).

2n + 2. $\vdash x \cdot p \equiv t(\bmod n) \supset \text{r.h.s.}$
⊃-intro. on 2n + 1.

2n + 3. $t \equiv 0(\bmod n) \ \& \ x \cdot p \equiv 0(\bmod n) \vdash x \cdot p \equiv t(\bmod n)$
Theorem 7.6 (iii).

⋮   ⋮

3n + 2. $t \equiv n - 1(\bmod n) \ \& \ x \cdot p \equiv n - 1(\bmod n) \vdash x \cdot p \equiv t(\bmod n)$
Theorem 7.6 (iii).

3n + 3. $\vdash \text{r.h.s.} \supset x \cdot p \equiv t(\bmod n)$
$2n + 3, \ldots, 3n + 2$.
∨-intro. and ⊃-intro.

7.10. Let l.h.s. and r.h.s. stand respectively for the left-hand side and the right-hand side of the equivalence. We first prove $\vdash$ l.h.s. $\supset$ r.h.s.

1. $x \equiv 0(\bmod N), x \cdot p_0 \equiv z_0(\bmod n_0) \vdash 0 \cdot p_0 \equiv z_0(\bmod n_0)$
(Theorem 7.6 (ix).

$x \equiv 0(\bmod N), x \cdot p_1 \equiv z_1(\bmod n_1) \vdash 0 \cdot p_1 \equiv z_1(\bmod n_1)$

$x \equiv 0(\bmod N), x \cdot p_2 \equiv z_2(\bmod n_2) \vdash 0 \cdot p_2 \equiv z_2(\bmod n_2)$

⋮

$k + 1.$ $\quad$ $x \equiv 0(\bmod N), x \cdot p_k \equiv z_k(\bmod n_k) \vdash 0 \cdot p_k \equiv z_k(\bmod n_k)$

$$\text{Theorem 7.6 (ix).}$$

$k + 2.$ $\quad$ $x \equiv 0(\bmod N), \text{l.h.s.} \vdash x \equiv 0(\bmod N) \& 0 \cdot p_0 \equiv z_0(\bmod n_0) \& \ldots$

$$\& 0 \cdot p_k \equiv z_k(\bmod n_k)$$

$$\text{\&-elim. and \&-intro.}$$

$k + 3.$ $\qquad\qquad\qquad\qquad$ $x \equiv 0(\bmod N) \vdash \text{l.h.s.} \supset \text{r.h.s.}$

$$\vee\text{-intro. and } \supset\text{-intro.}$$

$\vdots$

$N(k + 3).$ $\qquad\qquad$ $x \equiv N - 1(\bmod N) \vdash \text{l.h.s.} \supset \text{r.h.s.}$ $\quad$ similarly.

$$\vdash \text{l.h.s.} \supset \text{r.h.s.} \quad \vee\text{-elim. and}$$

$$\text{Theorem 7.6 (vi).}$$

To prove $\vdash \text{r.h.s.} \supset \text{l.h.s.}$

1. $\qquad$ $x \equiv 0(\bmod N), 0 \cdot p_0 \equiv z_0(\bmod n_0) \vdash x \cdot p_0 \equiv z_0(\bmod n_0)$

$$\text{Theorem 7.6 (ix).}$$

$x \equiv 0(\bmod N), 0 \cdot p_1 \equiv z_1(\bmod n_1) \vdash x \cdot p_1 \equiv z_1(\bmod n_1)$

$x \equiv 0(\bmod N), 0 \cdot p_2 \equiv z_2(\bmod n_2) \vdash x \cdot p_2 \equiv z_2(\bmod n_2)$

$\vdots$

$k + 1.$ $\quad$ $x \equiv 0(\bmod N), 0 \cdot p_k \equiv z_k(\bmod n_k) \vdash x \cdot p_k \equiv z_k(\bmod n_k)$

$$\text{Theorem 7.6 (ix).}$$

$k + 2.$ $\quad$ $x \equiv 0(\bmod N) \& 0 \cdot p_0 \equiv z_0(\bmod n_0) \& \ldots \& 0 \cdot p_k \equiv z_k(\bmod n_k)$

$$\vdash \text{l.h.s.}$$

$\vdots$ $\qquad\qquad\qquad\qquad\qquad\qquad$ \&-elim. and \&-intro.

$N(k + 2).$ $\quad$ $x \equiv N - 1(\bmod N) \& (N - 1)p_0 \equiv z_0(\bmod n_0) \& \ldots$

$$\& (N - 1)p_k \equiv z_k(\bmod n_k) \vdash \text{l.h.s.}$$

$$\text{similarly.}$$

$$\vdash \text{r.h.s.} \supset \text{l.h.s.} \qquad \vee\text{-elim.}$$

7.17. Let L be $\exists x(x \equiv z(\bmod n) \& a < x + t < b)$ and let R be
$[(a < t \& t + z < b) \vee E_1 \vee \ldots \vee E_n]$. Since $z < n, x \equiv z(\bmod n) \vdash \neg x < z$,
we do not need to consider the case when $x + k \cdot n = z$. The proof is as follows:

1. $\qquad\qquad\qquad$ $\vdash a < t \vee t < a'$ $\quad$ *139$_D$, *138a$_D$.

2. $\quad z < n, a < t \vdash x = k \cdot n + z \& a < x + t < b \supset a < t \& t + z < b$

$$\vdash k \cdot n + z + t < b \supset t + z < b.$$

3. $z < n, a < t \vdash L \supset R.$ $\qquad$ ∨-elim. and ∃-elim.

4. $\vdash x \equiv z (\bmod n) \sim x + t \equiv t + z (\bmod n)$

5. $\vdash x + t \equiv a + 1 (\bmod n) \lor \ldots \lor x + t \equiv a + n (\bmod n)$

6. $x + t \equiv a + 1 (\bmod n), a < x + t < b \vdash a + 1 < b$
$$\vdash c \equiv d (\bmod n)$$
$$\vdots \qquad \& \, d < n \supset d \leqslant c.$$

$5 + n.$ $x + t \equiv a + n (\bmod n), a < x + t < b \vdash a + n < b$

$6 + n.$ $\qquad L, t < a' \vdash R.$ $\quad$ 4, 5, 6 to $5 + n$, and
$\qquad\qquad\qquad\qquad\qquad\qquad\qquad$ ∨-elim.

$7 + n.$ $\qquad\qquad z < n \vdash L \supset R.$
$\qquad\qquad\qquad\qquad\qquad\qquad$ 1, 3, and $6 + n.$

Then, $\vdash z < n \supset (R \supset L)$ is a proof by cases.

7.18.

(i) For inequalities,

(a) if $m \cdot n$ is true we have:

1. $\qquad\qquad \vdash m < n \lor m \geqslant n$

2. $m = n + x \vdash 0 = ((n - m) - 1) + x)'$ $\qquad$ Axiom 19.

Note: $n - m$ is *not* an expression of our formal theory; we use it for ease of
reading instead of $0'' \cdots {}''(n-m \text{ times})$ where $n - m$ must be non-negative.

3. $\qquad\qquad \vdash \neg 0 = ((n - m) - 1) + x)'$ $\qquad$ Axiom 15.

4. $\qquad\qquad \vdash \neg m = n + x$ $\qquad$ 2, 3, and ¬-elim.

5. $\qquad\qquad \vdash \neg m \geqslant n$ $\qquad$ 4 and ∃-intro.

6. $\qquad\qquad \vdash m < n$ $\qquad$ 1, 5, and tautology.

(b) if $m < n$ is false we have instead:

1. $m + x' = n \vdash (m - n + x)' = 0$ $\qquad$ Axiom 14.

2. $\qquad\qquad \vdash \neg ((m - n) + x' = 0$ $\qquad$ Axiom 15.

3. $\qquad\qquad \vdash \neg m < n$ $\qquad$ 1, 2, ¬-elim., and ∃-intro.

(ii) For congruences consider only $k \equiv 0 (\bmod n)$, since
$\vdash k \equiv h (\bmod n) \sim k + (h \cdot (n - 1)) \equiv 0 (\bmod n).$

1. $\vdash k \equiv 0 (\bmod n) \sim k - n \equiv 0 (\bmod n)$ $\qquad$ Theorem 7.6, (iii),
$\qquad\qquad\qquad\qquad\qquad\qquad\qquad\qquad\qquad\qquad$ (iv), and (v).

$\vdots \qquad \vdots$

$m.$ $\vdash k - ((m - 1) \cdot n) \equiv 0 (\bmod n) \sim$
$$k - (m \cdot n) \equiv 0 (\bmod n) \, \& \, k - (m \cdot n) < n$$

(a) if $0 < k - (m \cdot n)$ is true continue with,

$m + 1.$ $\vdash 0 < k - (m \cdot n) < n$ part (i) (a)

$m + 2.$ $\vdash \neg(k - (m \cdot n) \equiv 0 \pmod{n}) \,\&\, 0 < k - (m \cdot n) < n)$

Theorem 7.6 (viii).

$m + 3.$ $\vdash \neg k \equiv 0 \pmod{n}$

$1, \ldots, m + 2,$ &-intro.,
and Theorem 7.6 (iii).

(b) if $0 < k - (m \cdot n)$ is false continue instead:

$m + 1.$ $\vdash \neg 0 < k - (m \cdot n)$ part (i) (b)

$m + 2.$ $\vdash 0 = k - (m \cdot n)$ $m + 1,$ Axiom 15.

$m + 3.$ $\vdash 0 \equiv 0 \pmod{n}$ Theorem 7.6 (i).

$m + 4.$ $\vdash k \equiv 0 \pmod{n}$

$m + 2, m + 3,$ and
Replacement Theorem,
then $1, \ldots, m$ and
Theorem 7.6 (iii).

Note: If $k < n$, then the lines $1, \ldots, m$ become, simply,
$\quad\quad \vdash k \equiv 0 \pmod{n} \sim k \equiv 0 \pmod{n} \,\&\, k < n.$

Chapter 8

8.1.   (i)   An infinite amount.

(ii)-(iv) Only you can check.

8.2.   If $r_1 \leqslant r_2$ the machine prints 0 in R_1 and stops. If $r_1 > r_2$ the machine adds 1 to the contents of R_1 and continues to repeat this — never stopping.

8.3.   The flow diagram and program for copying the contents of R_i into R_j, for $i \neq j$, are:

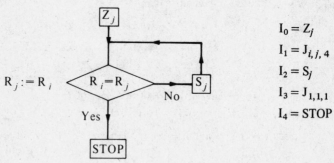

$I_0 = Z_j$

$I_1 = J_{i, j, 4}$

$I_2 = S_j$

$I_3 = J_{1, 1, 1}$

$I_4 = \text{STOP}$

8.4.
$$I_{13} = Z_j$$
$$I_{14} = J_{i,j,\,17}$$
$$I_{15} = S_j$$
$$I_{16} = J_{1,1,14}$$
$$I_{17} = S_i$$

8.5.   (i) No; (ii) Yes, R_3.

8.6.

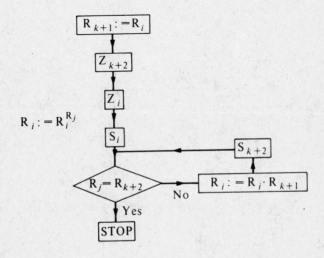

8.7.

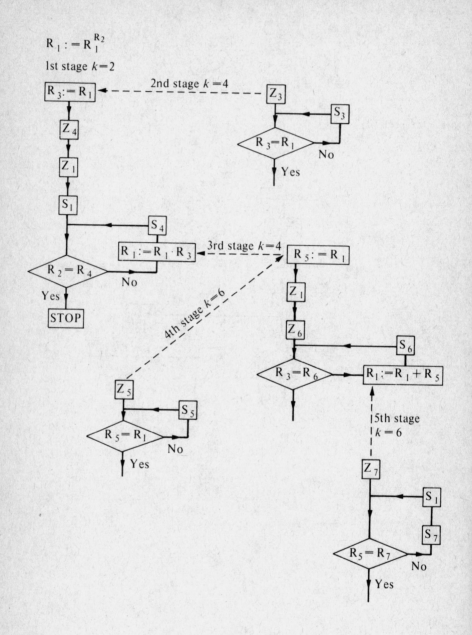

$R_1 := R_1^{R_2}$

1st stage $k=2$

$R_3 := R_1$ ← — — — 2nd stage $k=4$ — — — Z_3

Z_4

Z_1

S_1

S_4

$R_1 := R_1 \cdot R_3$ ← — — 3rd stage $k=4$

$R_2 = R_4$ No

Yes

STOP

Z_3 ← S_3

$R_3 = R_1$ No

Yes

$R_5 := R_1$

Z_1

Z_6 ← S_6

$R_3 = R_6$ → $R_1 := R_1 + R_5$

4th stage $k=6$

Z_5

S_5

$R_5 = R_1$ No

Yes

5th stage $k=6$

Z_7

S_1

S_7

$R_5 = R_7$ No

Yes

8.8.

$R_3 := R_1$
$k = 4$

$\left\{ \begin{array}{l} I_0 = Z_3 \\[4pt] I_1 = J_{3,1,4} \\[4pt] I_2 = S_3 \\[4pt] I_3 = J_{1,1,1} \end{array} \right.$

$I_4 = Z_4$

$I_5 = Z_1$

$I_6 = S_1$

$I_7 = J_{2,4,24}$

$R_1 := R_1 \cdot R_3$
$k = 4$

$\left\{ \begin{array}{l} I_8 = Z_5 \\[4pt] I_9 = J_{5,1,12} \\[4pt] I_{10} = S_5 \\[4pt] I_{11} = J_{1,1,9} \\[4pt] I_{12} = Z_1 \\[4pt] I_{13} = Z_6 \\[4pt] I_{14} = J_{3,6,22} \\[4pt] I_{15} = Z_7 \\[4pt] I_{16} = J_{5,7,20} \\[4pt] I_{17} = S_7 \\[4pt] I_{18} = S_1 \\[4pt] I_{19} = J_{1,1,16} \\[4pt] I_{20} = S_6 \\[4pt] I_{21} = J_{1,1,14} \end{array} \right.$

$I_{22} = S_4$

$I_{23} = J_{1,1,7}$

$I_{24} = STOP$

8.9. You have to find a program which computes it; there is no algorithm which will decide.

Chapter 9

9.1. Try it with a numerical example, say $3; 0! = 1$ is a special case, so check that; in a simple case like this numerical examples will be adequate checks, but there is no algorithm which will decide whether any given program is correct.

9.2.

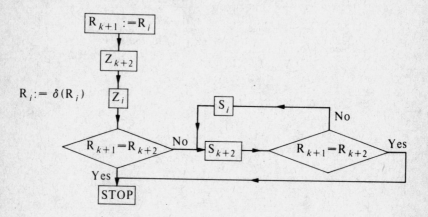

9.3.

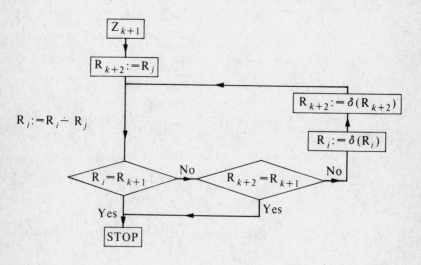

9.4.

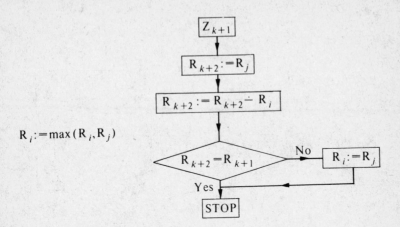

$R_i := \max(R_i, R_j)$

9.5.

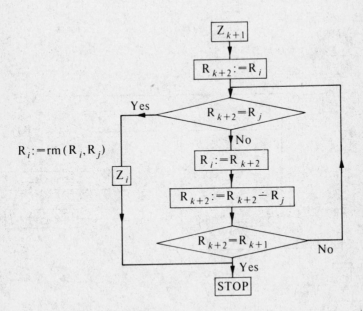

$R_i := \operatorname{rm}(R_i, R_j)$

9.6.

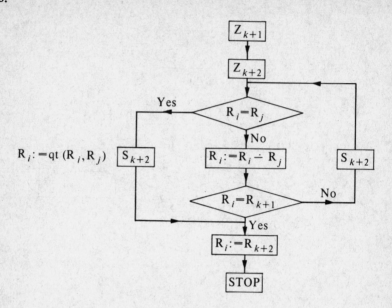

9.7.

$R_i := D(R_j)$

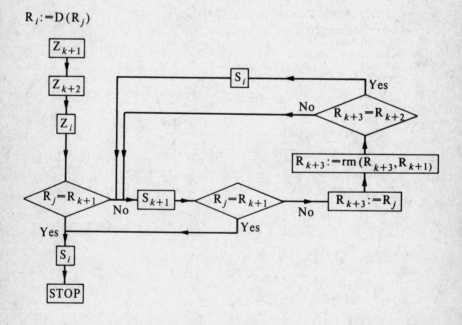

9.8.

$R_i := Pr(R_j)$

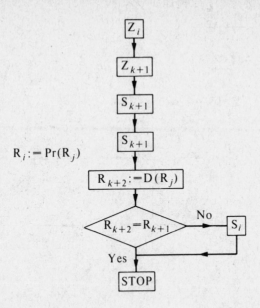

9.9. Assume there is such a program P_t and let R_h be the highest numbered register referred to in P_t. Clearly if $a_n > 0$ and $n > h$, P_t cannot check whether v is the gödel number of a term.

9.10.

$R_i := (R_j)_{R_q}$

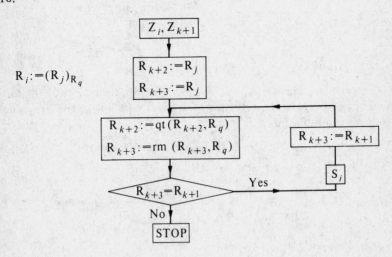

9.11.

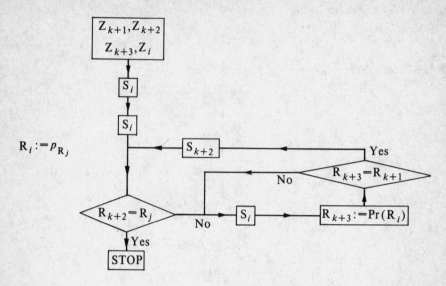

$R_i := p_{R_j}$

9.12.

$R_i := (R_j)_{p_{R_q}}$

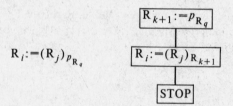

9.13.

(i) By (B) (p. 102).

(ii) It is necessary to ask if $y_{r_{k+3}+1} = 0$. Without this, the term $(a' + 0)'$, for example, would be reconstructed as far as $(a' + 0)$ in R_{k+1} and R_{k+1} would then be multiplied by $p_{m+2}^{x_{m+2}}$ again.

(iii)

(	a	+	(	0	+	(	a	·	b	'	)	'	)	'	)
10	1	10	8	2	8	6	3	6	4	5	6	7	8	9	10

(iv) (1) $y_{r_{k+3}} = 0$?

    (2) $y_{r_{k+3}-1} = 0$?

    (3) $y_{r_{k+3}+1} = 0$?

    (4) $y_m = 0$?

(v) (1) No. Stage II. (2) No. Stage II. (3) No. Stage V (without variables or 0, R_{k+1} remains $= 1$). (6) No. Stage V. (8) No. Stage IV. (10) No. (R_{k+1} will contain only a and b.) (11) No. Stage IV. (12) No. Stage IV. (15) No. Stage II. The others are correctly identified as terms.

Chapter 10

10.1. To show (*164) that $a = b$ is numeralwise expressed by a = b, let m, n be particular numbers.

(i) If $m = n$, clearly $\vdash m = n$, since m is $0'{\cdots}'^{m \text{ times}}$ and n is $0'{\cdots}'^{n \text{ times}}$.

(ii) If $m \neq n$, either $m < n$ or $m > n$. Suppose $m < n$. From Axiom 14 we have for arbitrary terms t, s $\vdash t' = s' \supset t = s$. Applying this m times gives

$$m = n \vdash 0 = r' \text{ (for some term r).}$$

Hence, by Axiom 15, etc., $\vdash \neg m = n$. Similarly for $m > n$.

To show (*165) that $a < b$ is numeralwise expressed by a < b, let m, n be natural numbers. We must show that:

$$\text{(i) if } m < n \text{ then } \quad \vdash \exists c(c' + m = n);$$

$$\text{(ii) if } m \not< n \text{ then } \quad \vdash \neg \exists c(c' + m = n).$$

(i) We illustrate the proof with $m = 2$ and $n = 5$. Clearly $\vdash 0''' + 0'' = (0''' + 0')'$ by Axiom 19. Then, $\vdash (0''' + 0')' = (0''' + 0)''$ by Axioms 19, 17, and 16. We can then get $\vdash (0''' + 0)'' = 0''''$ by Axioms 18, 17, and 16. Hence, $\vdash 0''' + 0'' = 0'''''$. By $\exists$-intro., $\vdash \exists c(c' + m = n)$.

(ii) We illustrate the proof with $m = 5$, $n = 2$. Suppose $\vdash \exists c(c' + m = n)$, then for some c, $\vdash c' + 0''''' = 0''$. By reasoning similar to (i), $\vdash c''''' = 0''$, thus $\vdash c'''' = 0$. But by Axiom 15, $\vdash \neg c'''' = 0$, therefore $\vdash \neg \exists c(c' + m = n)$. Similar reasoning applies for any numbers m, n.

10.2. We need a formula $A_f(x_1, x_2)$ of N such that if $m! = n$ then $\vdash A_f(m, n)$ and $\vdash \exists! x A_f(m, x)$.

10.3. All we need to show for numbers $m_1, \ldots, m_n, k$ is that if $f(m_1, \ldots, m_n) \neq k$ then $\vdash \neg F(m_1, \ldots, m_n, k)$. Assuming that $a = b$ is numeralwise expressible the result follows easily from $\vdash k \neq h$, $\vdash F(m_1, \ldots, m_n, h)$ and $\vdash \exists! b F(m_1, \ldots, m_n, b)$.

10.4. (*175). If m, n are particular numbers such that $m' = n$, clearly (i) $\vdash 0'^{(m' \text{ times})} = 0'^{(n \text{ times})}$ by *100. If t is a term which does not contain x, then $\vdash \exists! x(t = x)$ by *115 so, (ii) $\vdash \exists! b(m' = b)$.

(*176) Let k, m, n be particular numbers such that $m + n = k$. By Axiom 18

$\vdash m + 0 = m$, hence, by repeated applications of Axioms 19 and 17, and *102, (i) $\vdash m + n = k$ and (ii) $\vdash \exists! c(m + n = c)$, as in the previous example.

(*177) is similar to the previous example, using Axioms 20 and 21, instead of 18 and 19.

10.5. Our proofs that these functions are numeralwise representable in **N** all establish condition (ii)$'$ first, and then particularize for numerals m, n, etc.

10.6.

(i) $n = 2, h = 5, m = 3$.

(ii)   (a)

Moment	Instruction to be obeyed	Contents of R_1	R_2	R_3	...
0	0	3	2		
1	1	3	2	0	
2	2	3	2	0	
3	3	3	2	1	
4	4	4	2	1	
5	1	4	2	1	
6	2	4	2	1	
7	3	4	2	2	
8	4	5	2	2	
9	1	5	2	2	
10	5	5	2	2	

(ii) (b) $t^* = 10, \beta(c_0, d_0, t^*) = 5, \beta(c_1, d_1, 7) = 4$.

10.7.

(i) $\beta(c_0, d_0, t) = 2 \supset \{\beta(c_0, d_0, t + 1) = 3$

$\& \beta(c_3, d_3, t + 1) = 1 + \beta(c_3, d_3, t)$

$\& \forall i(\beta(c_i, d_i, t + 1) = \beta(c_i, d_i, t))\}$,

where $i = 1$ or $i = 2$.

(ii) $\beta(c_0, d_0, t) = 1 \supset \{(\beta(c_2, d_2, t) = \beta(c_3, d_3, t) \supset \beta(c_0, d_0, t + 1) = 5$

$\& (\beta(c_2, d_2, t) \neq \beta(c_3, d_3, t) \supset \beta(c_0, d_0, t + 1) = 2$

$\& \forall i(\beta(c_i, d_i, t + 1) = \beta(c_i, d_i, t))\}$,

where $1 \leqslant i \leqslant 3$.

(iii) $\beta(c_0, d_0, t^*) = 5$ & $\beta(c_1, d_1, t^*) = y$.

10.8.

(i) Assume $\beta(k_1, k_2, k_3) = 1 + \beta(m_1, m_2, m_3) = r + 1$ (for numbers k_i, m_j, r). By representability of the β-function,

$$\vdash \exists!w B(k_1, k_2, k_3, w) \qquad \text{and} \qquad \vdash B(k_1, k_2, k_3, r') \qquad (k)$$

and $\vdash \exists!w_1 B(m_1, m_2, m_3, w_1)$ and $\vdash B(m_1, m_2, m_3, r)$. $\quad(m)$

By *172 (which is A(r), A(s), $\exists!xA(x) \vdash r = s$, where r and s are terms which are free for x in A(x)), (k) and predicate logic, $B(k_1, k_2, k_3, w) \vdash r' = w$ and $B(m_1, m_2, m_3, w_1) \vdash r = w_1$. Hence, by &-elim. and Deduction Theorem $\vdash B(k_1, k_2, k_3, w)$ & $B(m_1, m_2, m_3, w_1) \supset w = w_1 + 1$. Thus, by $\forall$-intro.,

$$\vdash \forall w \forall w_1 [B(k_1, k_2, k_3, w) \ \& \ B(m_1, m_2, m_3, w_1) \supset w = w_1 + 1].$$

(ii) Assume $\beta(k_1, k_2, k_3) \neq 1 + \beta(m_1, m_2, m_3)$ and $\beta(m_1, m_2, m_3) = r$. By representability,

$$\vdash \exists!w B(k_1, k_2, k_3, w) \qquad \text{and} \vdash \neg B(k_1, k_2, k_3, r')$$

$$\vdash \exists!w_1 B(m_1, m_2, m_3, w_1) \quad \text{and} \vdash B(m_1, m_2, m_3, r).$$

Assume $\vdash \forall w \forall w_1 [(B(k_1, k_2, k_3, w)$ & $B(m_1, m_2, m_3, w_1)) \supset w = w_1 + 1]$ and (similarly to part (i)) it is easy to prove $\vdash B(k_1, k_2, k_3, r')$. Hence, by contradiction,

$$\vdash \neg \forall w \forall w_1 [B(k_1, k_2, k_3, w) \ \& \ B(m_1, m_2, m_3, w_1) \supset w = w_1 + 1].$$

10.9. $n = 2, h = 24, m = 7$.

$$A(t) \sim A_0(t) \ \& \ \dots \ \& \ A_{23}(t) \ \& \ [B(c_0, d_0, t, r) \supset r < 24].$$

Each $A_j(t)$ expresses I_j of the program for computing a^b.

10.10. We suppose that there is an R-computation of $f(k_1, \dots, k_n) = k_{n+1}$. Find numbers c_0, d_0, to code the sequence of instructions obeyed; find c_i, d_i to code the contents of R_i for $1 \leqslant i \leqslant m$; find t^*, etc. By the representability of the β-function, and other results we easily get

$$\vdash_N S(k_1, \dots, k_n) \ \& \ \forall t(t < t^* \supset A(t)) \ \& \ H(t^*, k_{n+1}).$$

Hence, by $\exists$-intro. applied repeatedly, $\vdash_N F_{P_{h+1}}(k_1, \dots, k_n, k_{n+1})$.

10.11. Abbreviate $F_{P_{h+1}}(k_1, \dots, k_n, y)$ to $F(y)$. We know that $\vdash F(k_{n+1})$ (by Exercise 10.10). By the definition of $\exists!$ (p. 113), we need to show that $\vdash_N F(y) \supset y = k_{n+1}$. Then by $\forall$-intro., &-intro., and, finally, $\exists$-intro., we have (ii).

10.12. In view of the very full answer given to 10.8, the following outline should suffice.

(I) *Basis* $\vdash \phi \supset S(k_1,\ldots,k_n)$, by &-elim.

$\vdash S(k_1,\ldots,k_n) \supset B(c_0, d_0, 0, 0)$, by &-elim.

$\vdash B(c_0, d_0, 0, 0)$, by representability of β-function.

Thus, $\vdash \phi \supset [B(c_0, d_0, 0, 0)\ \&\ B(c_0, d_0, 0, 0)]$.

Similarly, for $i, 0 < i \leqslant m$.

Thence, by *strong* representability of β-function, applied to $B(c_0, d_0, 0, 0)$, etc. and by the method of 10.8, we get

$$\vdash \phi \supset \forall w \forall w_1 [B(c_i, d_i, 0, w)\ \&\ B(c_i, d_i, 0, w_1) \supset w = w_1].$$

(II) *Induction step* Suppose, for $n < t^*$, that

$$\vdash \phi \supset \forall w \forall w_1 [B(c_i, d_i, n, w)\ \&\ B(c_i, d_i, n, w_1) \supset w = w_1]$$

and $\vdash B(c_0, d_0, n, j)$. Hence, $\vdash \phi \supset B(c_0, d_0, n, j)$.

There are the three cases to consider: I_j is either (1) S_k; (2) Z_k; or (3) $J_{k,r,q}$.

Cases (1) and (2): clearly, $\vdash \phi \supset B(c_0, d_0, n + 1, j + 1)$, etc., so

$$\vdash \phi \supset \forall w \forall w_1 [B(c_i, d_i, n + 1, w)\ \&\ B(c_i, d_i, n + 1, w_1) \supset w = w_1].$$

Case (3): by the induction hypothesis, etc., either

(i) $\phi \vdash \forall w \forall w_1 [B(c_k, d_k, n, w)\ \&\ B(c_r, d_r, n, w_1) \supset w = w_1]$

or (ii) $\phi \vdash \neg \forall w \forall w [B(c_k, d_k, n, w)\ \&\ B(c_r, d_r, n, w_1) \supset w = w_1]$.

If (i), then $\phi \vdash B(c_0, d_0, n + 1, q)$, etc., and we already know that $\vdash B(c_0, d_0, n + 1, q)$, and so on. If (ii), then $\vdash B(c_0, d_0, n + 1, j + 1)$, etc. Thence, as in the basis, we get

$$\vdash \phi \supset \forall w \forall w_1 [B(c_i, d_i, n + 1, w)\ \&\ B(c_i, d_i, n + 1, w_1) \supset w = w_1].$$

10.13.

(i)   By Exercise 10.12.

(ii) &-elim. in both cases.

(iii) Assume $t^* < t^*$ and contradiction results from (i) and (ii).

(iv) &-elim.

(v)   From the assumptions that a genuine computation exists, that $\vdash a \leqslant n \supset a = 0 \lor a = 1 \lor a = 2 \lor \ldots \lor a = n$ and that $\beta(c, d, i)$ is representable.

(vi)   By *139, $\vdash a < b \lor a = b \lor a < b$.

(vii) Because $\vdash B(c_1, d_1, t^*, k_{n+1})$.

(viii) By $\vdash \exists! w B(c_1, d_1, t^*, w)$.

Chapter 11

11.1.

(i)   For every a, b either b is not the gödel number of a proof of the formula
       $A(a)$ or there is a number c, less than or equal to b, which is the gödel
       number of a proof of $\neg A(a)$. Or, very briefly: for every a, if $A(a)$ is
       provable, so is $\neg A(a)$.

(ii)   No, if N is consistent. Let a be the gödel number of the formula c = c.
       We know by Exercise 5.9, etc., that $\vdash_N a = a$.

11.2.

(i)   For every b, either b is not the gödel number of a proof of the formula
       $G(g)$, or there is a number c less than or equal to b which is the gödel
       number of a proof of $\neg G(g)$. Or, very briefly: if $G(g)$ is provable, so
       is $\neg G(g)$.

(ii)   Not so easy to say, but the proof of the Gödel–Rosser Theorem will
       convince you that it is true.

11.3. Assume that n is the gödel number of a formula with exactly the one free
variable x, and let the registers have the following initial contents: $R_1 = 1$;
$R_2 = gn(x)$; $R_3 = n$; $R_4 =$ the largest prime factor of n; $R_5 = R_6 = R_7 = R_8 = 0$.
The answer is in R_1.

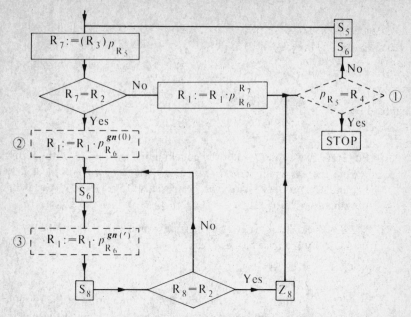

The abbreviations are self-explanatory; they expand as follows.

① Put p_{R_5} in a suitable, unused, register R_k. Is $R_k = R_4$? We suppose that the gödel numbers of the formal symbols 0 and $'$ are already in some suitable register, R_z and R_s.

② Multiply R_1 by $p_{R_6}^{R_z}$ and put the answer in R_1.

③ Similarly, with R_s in place of R_z.

11.4. $Q(a)$ says: 'If a is the gödel number of a formula $A(x)$ with exactly one free variable x, $A(a)$ is not provable. Clearly, it is false for x = x and true for some formulas.

$Q(q)$ says: '$Q(q)$ is not provable', *or*, 'I am not provable'. The proof shows that $Q(q)$ is true.

11.5. Assume $C_{\text{Prov}_T}(n)$ is representable in **T** by the formula $B(a, b)$. If $\text{Prov}_T(n)$ is *true*, $C_{\text{Prov}_T}(n) = 0$. Hence, $\vdash_T B(n, 0)$. If $\text{Prov}_T(n)$ is false, $C_{\text{Prov}_T}(n) = 1$. It follows that $\vdash_T B(n, 1)$, hence by uniqueness, $\vdash_T \neg B(n, 0)$.

11.6. If **T** is consistent and every total R-computable function is representable in **T**, Diag(n) is. By Theorem 11.4, $\text{Prov}_T(n)$ is not expressible in **T**. If $C_{\text{Prov}_T}(n)$ is R-computable then it is representable in **T**. In that case, $\text{Prov}_T(n)$ is expressible in **T**, which is impossible.

11.7. Assume **N** is consistent. We know that every total R-computable function is representable in **N** (Theorem 10.2). By Theorem 11.6, $C_{Prov_N}(n)$ is not R-computable. Therefore, by definition 11.3, **N** is R-undecidable.

11.8. Let **N⁺** be any consistent extension of **N**. Since **N⁺** is consistent and every total R-computable function is representable in **N⁺**, $C_{Prov_N^+}(n)$ is not R-computable by Theorem 11.6. Hence **N⁺** is R-undecidable.

11.9. Immediate consequence of previous theorem.

11.10. If we run through our enumeration of theorems, we will eventually encounter either A or ⌐A. More precisely, we programme an R-machine to decide for successive natural numbers n whether n is the gödel number of a proof in **T** and, if so, whether the last formula in the proof is A or ⌐A.

11.11. Let **K** be a consistent, axiomatizable extension of **N**. Then **K** is both R-undecidable by Theorem 11.8, and R-decidable by Theorem 11.10.

11.12. Since **A** is a consistent extension of **N**, the result is immediate from Theorem 11.8.

11.13. Since **A** is a consistent, complete extension of **N**, the result is immediate from Theorem 11.11.

11.14. We assume **A** is consistent. If $Tr(n)$ then, since n is the gödel number of an axiom, $Prov_A(n)$. If not $Tr(n)$, since we assume **A** is consistent, ⌐$Prov_A(n)$. Hence, $Tr(n)$ iff $Prov_A(n)$. If **A** is consistent, since all total R-computable functions are representable in **A**, by Theorem 11.4 we have $Prov_A(n)$ is not expressible in **A**. Hence, $Tr(n)$ is not expressible in **A**.

Chapter 12

12.1. Proof is by induction in the metalanguage. We first show that
$\vdash_{Q^+} a \leqslant 0 \supset a = 0$.

1.	$a \leqslant 0 \vdash a = 0 \lor a < 0$	definition.
2.	$a < 0 \vdash \exists c(c' + a) = 0)$	definition.
3.	$\vdash a = 0 \lor \neg a = 0$	
4.	$a = 0 \vdash c' + a = c' + 0$	Replacement Theorem.
5.	$a = 0 \vdash c' + a = c'$	4, Axiom 22.
6.	$c' + a = 0, a = 0 \vdash c' = 0$	5.

7.	$a = 0 \vdash \neg c' + a = 0$	6, Axiom 20, $\neg$-intro.
8.	$\neg a = 0 \vdash \exists b(a = b')$	Axiom 21.
9.	$a = b' \vdash c' + a = c' + b'$	Replacement Theorem.
10.	$a = b' \vdash c' + a = (c' + b)'$	9, Axiom 23.
11.	$c' + a = 0, a = b' \vdash (c' + b)' = 0$	10.
12.	$a = b' \vdash \neg c' + a = 0$	11, Axiom 20, $\neg$-intro.
13.	$\neg a = 0 \vdash \neg c' + a = 0$	8, 12, $\exists$-elim.
14.	$\vdash \neg c' + a = 0$	3, 7, 13, $\vee$-elim.
15.	$\vdash \neg a < 0$	2, 14, $\neg$-intro, $\exists$-elim.
16.	$a \leqslant 0 \vdash a = 0$	1, 15.

For the induction step we proceed as follows:

17.	$a \leqslant n' \vdash a = n' \vee a < n'$	definition.
18.	$a < n' \vdash \exists c(c' + a = n')$	definition.
19.	$\neg a = 0 \vdash \exists b(a = b')$	Axiom 21.
20.	$c' + a = n', a = b' \vdash (c' + b)' = n'$	Axiom 23.
21.	$c' + a = n', a = b' \vdash c' + b = n$	20, Axiom 19.
22.	$c' + a = n', a = b' \vdash \exists b(a = b' \,\&\, c' + b = n)$	21, $a = b' \vdash a = b'$ and $\exists$-intro.
23.	$a < n', \neg a = 0 \vdash \exists b(a = b' \,\&\, c' + b = n)$	19, 18, $\exists$-elim., twice.
24.	$a < n' \vdash a = 0 \vee \exists b(a = b' \,\&\, c' + b = n)$	23, $\vdash (\neg A \supset B) \sim (A \vee B)$.
25.	$a = b' \,\&\, c' + b = n \vdash b \leqslant n$	&-elim., $\exists$-intro., definition of $<$, $\vee$-intro.
26.	$b \leqslant n \vdash b = 0 \vee b = 1 \vee \dots \vee b = n$	Hypothesis of the induction.
27.	$b = 0, a = b' \vdash a = 0'$	
28.	$b = 1, a = b' \vdash a = 1'$	
$\vdots$	$\vdots$	
$27 + n.$	$b = n, a = b' \vdash a = n'$	
$28 + n.$	$a = b' \,\&\, c' + b = n \vdash a = 1 \vee \dots \vee a = n'$	25, 26, $27 - 27 + n$, $\vee$-intro., and $\vee$-elim.
$29 + n.$	$a < n' \vdash a = 0 \vee a = 1 \vee \dots \vee a = n'$	24, 25, $\exists$-elim., $\vee$-elim., and $\vee$-intro.

$30 + n.$ $a = n' \vdash a = 0 \lor a = 1 \lor \ldots \lor a = n'$ $\lor$-intro.

$31 + n.$ $a \leqslant n' \vdash a = 0 \lor a = 1 \lor \ldots \lor a = n'$ $29 + n, 30 + n, 17,$
                                                                                      $\lor$-elim.

This establishes that for any numeral n

$$\vdash_{Q^+} a \leqslant n \supset a = 0 \lor a = 1 \lor \ldots \lor a = n.$$

12.2.

(i) 1. $\vdash a \leqslant n \supset a = 0 \lor a = 1 \ldots \lor a = n$
                                                                     Lemma 12.3 (i).

     2. $a = 0, n' + 0 = n' \vdash \exists c(c' + a = n')$ Replacement Theorem,
                                                                     $\exists$-intro.
          $\vdots$ $\vdots$

$n + 2.$ $a = n, 0' + n = n' \vdash \exists c(c' + a = n')$

$n + 3.$ $\vdash_{Q^+} a \leqslant n \supset a \leqslant n'.$ $1, 2, \ldots, n + 2, \lor$-elim.,
                                                                     $\cdot$ representability of
                                                                     $r + s = t$, and $\exists$-intro.

(ii) By repeated applications of Axioms 22, 23, and 16, $\vdash_{Q^+} a' + n = a + n'$.
For example, $a' + 0'' = (a' + 0')' = (a' + 0)'' = a'''$ and
$a + 0''' = (a + 0'')' = (a + 0')'' = (a + 0)''' = a'''$.

(iii) 1. $n < a \vdash \exists c(c' + n = a)$ definition.

      2. $\vdash \neg c' = 0$ Axiom 20.

      3. $c' + n = a \vdash \exists b(b' + n = a)$ 2, Axiom 21, $\exists$-intro.,
                                                                   $\exists$-elim.

      4. $b' + n = a \vdash b + n' = a$ (ii)

      5. $\vdash b = 0 \lor \neg b = 0$ $\vdash A \lor \neg A.$

      6. $b' + n = a, b = 0 \vdash 0 + n' = a$ (ii), Replacement
                                                                   Theorem.

      7. $b' + n = a, b = 0 \vdash n' = a$ 6, (ii).

      8. $\neg b = 0 \vdash \exists d(b = d')$ Axiom 21.

      9. $b + n' = a, b = d' \vdash d' + n' = a$ Replacement Theorem.

     10. $\vdash n' \leqslant a$ 9, 8, 6, 5, 4, 3, 1,
                                                                   $\exists$-intro., $\exists$-elim.,
                                                                   $\lor$-elim., $\exists$-elim., $\exists$-elim.

Hence $\vdash_{Q^+} n < a \supset n' \leqslant a.$

To prove Lemma 12.3 (ii), use induction in the metalanguage.

We first show that $\vdash_{Q^+} a \leqslant 0 \vee 0 \leqslant a$.

1.	$a = 0 \vdash a \leqslant 0 \vee 0 \leqslant a$	definition, $\vee$-intro.
2.	$a = 0 \vdash \exists b(a = b')$	Axiom 21.
3.	$a = b' \vdash a = b' + 0$	Axiom 22.
4.	$a = b' \vdash \exists c(c' + 0 = a)$	3, $\exists$-intro.
5.	$\neg a = 0 \vdash \exists c(c' + 0 = a)$	2, 4, $\exists$-elim.
6.	$\neg a = 0 \vdash 0 \leqslant a \vee a \leqslant 0$	5, definition, $\vee$-intro.
7.	$\vdash 0 \leqslant a \vee a \leqslant 0$	1, 6, $\vdash A \vee \neg A$, $\vee$-elim.

For the induction step assume $\vdash_{Q^+} a \leqslant n \vee n \leqslant a$.

8.	$\vdash a \leqslant n \vee n \leqslant a$	
9.	$a \leqslant n \vdash a \leqslant n' \vee n' \leqslant a$	(i) and $\vee$-intro.
10.	$n \leqslant a \vdash a = n \vee n' \leqslant a$	(iii).
11.	$a = n \vdash a \leqslant n$	$\vee$-intro.
12.	$a \leqslant n \vdash a \leqslant n'$	(i).
13.	$n \leqslant a \vdash a \leqslant n' \vee n' \leqslant a$	12, 11, 10, $\vee$-intro., $\vee$-elim.

12.3.

(i) Diag(n) is R-computable, by Theorem 11.3. So, by Theorem 12.4, Diag(n) is numeralwise representable in Q^+.

(ii) Q^+ is a first-order formal theory with the same symbols as N and for which the usual properties of identity are provable. Hence, the result is immediate from part (i) and Theorem 11.4.

(iii) By definition 11.3 and Theorem 11.6.

(iv) Let Z be a consistent extension of Q^+. Since Z is consistent and every total R-computable function is representable in Z, $C_{Prov_Z}(n)$ is not R-computable by Theorem 11.6. Hence Z is R-undecidable.

12.4.

(i) Since Axiom (schema) 13 represents a different proper axiom for every different $A(x)$, N has denumerably many proper axioms.

(ii) No. Our answer does not show that *there is no* finite set of proper axioms from which all the theorems of N are derivable, although this is in fact true.

12.5.

 (i)   **N** is *not* equivalent to **D** plus the *two* axioms for '·', Axioms 20 and 21;
       N also has all the *logical* axioms containing '·'.

 (ii) **N**.

 (iii) If **N** is consistent, **N** and $\mathbf{N}^+$ are compatible. $\mathbf{N}^+$ and $\mathbf{N}^{\neg}$ are *not* compatible.

12.6. By definition 12.1, there is a finite set of formulas $E_1, \ldots, E_n$ such that

$$\vdash_{\mathbf{T}_2} A \text{ iff } E_1, \ldots, E_n \vdash_{\mathbf{T}_1} A.$$

By the Deduction Theorem,

$$\vdash_{\mathbf{T}_1} E_1 \supset (E_2 \supset \ldots \supset (E_n \supset A) \ldots) \text{ iff } \vdash_{\mathbf{T}_2} A.$$

Hence, if $\mathbf{T}_1$ is R-decidable, so is $\mathbf{T}_2$. Hence, if $\mathbf{T}_2$ is R-undecidable, so is $\mathbf{T}_1$.

12.7. If **T** is a consistent extension of $\mathbf{Q}^+$, then, since $\mathbf{Q}^+$ is essentially R-undecidable (Theorem 12.5 (iv)), **T** is R-undecidable. If **T** is a first-order subtheory of $\mathbf{Q}^+$, then, since $\mathbf{Q}^+$ is finitely axiomatizable, $\mathbf{Q}^+$ is a finite extension of **T**. By Theorem 12.6, **T** is R-undecidable.

12.8. Immediate by Theorem 12.7.

12.9. Suppose there is an R-machine M_ν which can decide, for any formula of predicate logic, whether it is valid. Let A be a formula of **Pr** (same symbols as **N**). Since $\vdash_{\mathbf{Pr}} A$ iff A is logically valid, if M_ν exists, **Pr** is R-decidable. This contradicts Theorem 12.8.

12.10. $2h + 1, 2h + 2, 2h + 3$ are relatively prime, and, therefore, by the Chinese Remainder Theorem (Theorem 1.11), c is *unique*.

12.11.

R_3	R_4
0	3^2
1	3^2
2	3^2
3	$3^2 \cdot 7^1$
1	$3^2 \cdot 7^1$
2	$3^2 \cdot 7^1$
3	$3^2 \cdot 7^2$
1	$3^2 \cdot 7^2$
4	$3^2 \cdot 7^2$

12.12.

(i)   Suppose f(a) is computable and let M_p be an R-machine which computes it. Thus f(a) = $\phi(p, a)$, for all a. Now, consider f(p). By definition,

$$f(p) = \phi(p, p) + 1$$

as we have assumed $\phi(p, p)$ is defined. But, as M_p computes f(a) for all a, we have

$$f(p) = \phi(p, p).$$

So, by contradiction, f(a) is *not* computable.

(ii) No. There are denumerably many R-machines which are equivalent to M_p, and there is no algorithm for deciding when two R-machines are equivalent (i.e. compute the same values for the same arguments).

12.13. Let *Pr*(a, b) be the predicate 'b is the gödel number of a proof in **N** of the formula whose gödel number is a'. It is easy to believe that *Pr*(a, b) is R-decidable, so let M_p print 0 if b is a proof of a and 1 if b is not a proof of a. We now use M_p to construct another machine M_a. With a in R_1 and 1 in R_2, proceed as follows:

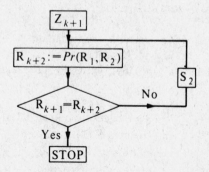

Then, M_a halts iff a is provable. If the halting problem is solvable, then **N** is R-decidable. Hence, if **N** is R-undecidable, the halting problem is unsolvable.

BIBLIOGRAPHY

Arrow, K. J. (1951). *Social Choice and Individual Values* (Cowles Foundation Monographs, 12) Wiley, New York. [2nd edition 1963.]

Bell, J. L. and Machover, M. (1977). *A Course in Mathematical Logic* North Holland, Amsterdam.

Boolos, G. S. and Jeffrey, R. C. (1974). *Computability and Logic* Cambridge University Press. [2nd edition 1980.]

Cantor, G. (1895-7). Beiträge zur Begründung der transfiniten Mengenlehre. *Mathematische Annalen*, vol. 46 (1895), pp. 481-512 and vol. 49 (1897), pp. 207-46. English translation by P. E. B. Jourdain (1915) called *Contributions to the founding of the theory of transfinite numbers*. Open Court, Chicago and London.

Church, A. (1936*a*). An unsolvable problem of elementary number theory. *American journal of mathematics*, vol. 58, pp. 345-63. [Also in Davis (1965).]

— (1936*b*). A note on the Entscheidungsproblem. *Journal of Symbolic Logic*, vol. 1, pp. 40-1. Correction, ibid., pp. 101-2. [Also in Davis (1965).]

— (1956). *Introduction to mathematical logic*, vol. I. Princeton University Press, N.J.

Crossley, J. N., Ash, C. J., Brickhill, C. J., Stillwell, J. C., and Williams, N. H. (1972). *What is Mathematical Logic?* Oxford University Press.

Davis, M. (ed.) (1965). *The Undecidable: Basic Papers on Undecidable Propositions, Unsolvable Problems and Computable Functions*. Raven Press, New York.

Dummett, M. A. E. (1963). The Philosophical Significance of Gödel's Theorem. *Ratio*, Vol. V, December 1963, No. 2, pp. 140-55.

Fraenkel, A. A. (1953). *Abstract Set Theory*. North Holland, Amsterdam. [2nd edition 1961. 3rd edition 1966. 4th edition 1976.]

Frege, G. (1879). *Begriffsschrift, eine der arithmetischen nachgebildete Formelsprache des reinen Denkens*. Nebert, Halle.

— (1884). *Die Grundlagen der Arithmetik, eine logisch-mathematische Untersuchung über den Begriff der Zahl*. Koebner, Breslau. [English translation by J. L. Austin (1950). *The Foundations of Arithmetic. A logico-mathematical enquiry into the concept of number*. Blackwell, Oxford.]

— (1893). *Grundgesetze der Arithmetik, begriffsschriftlich abgeleitet*, vol. I. H. Pohle, Jena.

— (1903). Ibid., vol. 2. [Partial translation of (1893) and (1903) in *The Basic Laws of Arithmetic: Exposition of the System*, edited by M. Furth, University of California Press, 1964.]

Gödel, K. (1931). Über formal unentscheidbare Sätze der Principia Mathematica und verwandter System I. *Monatshefte für Mathematik und Physik*, vol. 38, pp. 173-98. [Translated in Davis (1965).]

— (1944). Russell's mathematical logic. In *The Philosophy of Bertrand Russell*, edited by P. A. Schlipp, Northwestern University, Evanston and Chicago, pp. 123–53. Also in *Philosophy of Mathematics: Selected Readings*, edited by P. Benacerraf and H. Putnam (1964), Blackwell, Oxford, pp. 211–32.

Hilbert, D. and Bernays, P. (1934). *Grundlagen der Mathematik*, vol. 1. Springer Verlag, Berlin. [Reprinted J. W. Edwards, Ann Arbor, Michigan, 1944.]

Kleene, S. C. (1952). *Introduction to Metamathematics*. North Holland, Amsterdam.

— (1967). *Mathematical Logic*. Wiley, New York.

Mendelson, E. (1964). *Introduction to Mathematical Logic*. Van Nostrand, Princeton.

Nagel, E. and Newman, J. R. (1959). *Gödel's Proof*. Routledge and Kegan Paul, London.

Presburger, M. (1930). Über die Vollständigkeit eines gewissen Systems der Arithmetik ganzer Zahlen, in welchem die Addition als einzige Operation hervortritt. *Sprawozdanie z I Kongresu Matematykow Krajow Slowianskich (Comptes-rendus du I Congres des Mathematiciens des Pays Slaves)*, Warszawa 1929, Warsaw 1930, pp. 92–101, 395. [The proof is in Hilbert and Bernays (1934).]

Robinson, R. M. (1950). Abstract. An essentially undecidable axiom system. *Proceedings of the International Congress of Mathematicians (Cambridge, Mass., USA, August 30–September 6 1950)*, 1952, vol. 1, pp. 729–30.

Rosser, J. B. (1936). Extensions of some theorems of Gödel and Church. *Journal of Symbolic Logic*, vol. 1, pp. 87–91.

— (1939). An Informal Exposition of Proofs of Gödel's Theorems and Church's Theorem. *Journal of Symbolic Logic*, Vol. 4, No. 2, pp. 53–60.

— (1953). *Logic for Mathematicians*. McGraw-Hill, New York.

Shepherdson, J. C. and Sturgis, H. E. (1963). Computability of Recursive Functions. *Journal of the Association of Computing Machinery* Vol. 10, pp. 217–55.

Stewart, B. M. (1964). *Theory of Numbers* (2nd edition). Macmillan, New York.

Tarski, A. (1936). Der Wahrheitsbegriff in den formalisierten Sprachen. *Studia Philosophica* vol. 1, pp. 261–405. [English translation in *Logic, Semantics and Metamathematics: Papers from 1923 to 1938 by Alfred Tarski*. Translated by J. H. Woodger. Oxford University Press.]

—, Mostowski, A., and Robinson, R. M. (1953). *Undecidable Theories*. North Holland, Amsterdam.

Whitehead, A. N. and Russell, B. (1910–13). *Principia Mathematica* Vol. 1, 1911; Vol. 2, 1912; Vol. 3, 1913; Cambridge University Press.

INDEX